THE PROPERTY FINANCE SOURCEBOOK 2008

Compiled and edited by **Richard von Götzen**
Additional Research by **Balazs Gal**

2008

Routledge
Taylor & Francis Group
LONDON AND NEW YORK

First published 1994 by Estates Gazette
Second edition published 1999
Third edition published 2000
Fourth edition published 2002

Published 2014 by Routledge
2 Park Square, Milton Park, Abingdon, Oxon OX14 4RN
605 Third Avenue, New York, NY 10017, USA

Routledge is an imprint of the Taylor & Francis Group, an informa business

Cover design by Rebecca Caro, image supplied by Alamy.com
Typeset in Futura Light BT 10/12 by Amy Boyle, Rochester, Kent

ISBN 13: 978-0-7282-0460-7 (hbk)

DOI: 10.4324/9781315040561

Contents

Introduction
by Richard von Götzen

In your hands you have the latest edition of *The Property Finance Sourcebook*. The last edition was printed six years ago. The reason we left out the last few years was that we thought that, with the advent of so many online sources, there was no longer any real need for a hard copy book. We were wrong.

Over the past two years, we have received so many requests to bring the book up to date that we agreed to do so. Once again, it remains for me to thank all the many principal lenders for their continued support and assistance, without which it would be impossible to produce *The Property Finance Sourcebook*.

It is only as a result of the ongoing help and cooperation of these lenders that *The Property Finance Sourcebook* has been able to establish itself as the prime reference for the funding of commercial and residential investment properties, as well as for development finance in the UK. Now that we have seen that there is still so much demand, it is our goal to continue to build on this. In order to help us do so, we very much welcome any feedback from the users of the directory, for whom, after all, it has been produced. Comments and suggestions should be addressed to the editor.

The Market

Much has changed over the past three years. There have been a lot of new entrants and some old established companies have been taken over. In a property bull market it is not surprising to note that many new and existing entrants have become far more aggressive and flexible with their lending.

In the last edition, there were not that many institutions who would finance speculative development, but that has now changed. The maximum loan to value has also increased considerably. These changes mean that the borrower has far more choice and can probably pick up a much better deal.

Another change has been the consolidation in the property lending business. More than three-quarters of all UK commercial property debt is held by just 12 banks, while the largest half a dozen banks control almost 60%.

Commercial property lending has been increasing by between 25% and 30% a year since 2003 and while most of the funding had been for investment this changed in 2003 and 2004 and over half of all lending was for refinancing.

Two-thirds of all commercial lending came from UK banks and just over 10% from the German banks. However, when it comes to securitisation of debt this picture reverses, with over 80% carried out by German and US banks.

Average loan to value ratios have risen from 70% in 1998 to around 85% in 2007 while, at the same time, interest to income cover has fallen to the lowest level since we first published our sourcebook in 1994.

How to Use the Sourcebook

Our objective remains to provide information on all known lenders in the UK commercial property industry, together with as detailed a loan criteria as possible. Readers should, however, continue to view and use the sourcebook primarily as a broad guide because lenders will always evaluate propositions on a case-by-case basis.

Borrowers should, in the first instance, telephone the named contact at the lending institution and provide outlined particulars of the property to be financed. Only once this outline has been met with a positive response will it usually be necessary to send a complete package of information to then enable the lender to make a written conditional proposal or offer. This offer is always subject to a satisfactory independent valuation being obtained, usually at the expense of the borrower, and is further subject to contract.

A good, clear and concise presentation of the proposal requiring funding is vital, especially as most lenders receive a large number of competing propositions each day. The feedback received from lenders continues to underline this aspect, which still seems the most neglected by borrowers. It is the best presented and easiest to understand proposals that are most likely to receive the earliest and most favourable consideration. That said, written proposals should be as brief as possible and to the point.

Initially, the most important information that the lending institutions will need to know and will require verification for are:

1. The status and background of the borrower. In today's market there is no question that the track record of the borrower remains vital.
2. Brief details and a description of the property or project being proposed.
3. The extent of loan security afforded by the covenant, together with the length and the terms of such lease where applicable.
4. The security and revenue cash flow of the property or project, the potential for upside return and sensitivity to risk.

The mainstream lenders still remain driven by covenant strength and therefore as much relevant information as possible (the covenant's standing etc) should be included.

In order to secure a written quotation or offer of financing from a lending institution, the following should be sent with a proposal (not in any order of importance).

The Borrower

5. Audited accounts for the last three years and a statement of assets and liabilities. In the absence of audited accounts, certified or management accounts may in some cases be acceptable.
6. A curriculum vitae for all the principals of the borrowing company or for the borrower where appropriate.
7. A business plan if relevant.

The Property

8. Full details of the tenants, tenancies, rental income and the latest annual report and accounts for the tenants where available.
9. A copy of a recent valuation if available.
10. A photograph of the property.
11. The amount of loan requested, period for which required and the type of loan, ie, fixed, bullet or variable interest payments.

Developments

12. A full business plan. (This is absolutely vital.)
13. Complete details of the proposed development to include specifications, proposed contractor, architect and developer, together with cash flows, projections and details of any pre-lets or pre-sales.
14. A market analysis of occupier demand and supply for the type of property proposed in the geographical locality.

All in all, the increased number of lenders, the expansion of the sectors and the improved criteria on offer once again afford the borrower the best and most competitive opportunity for funding for many years.

We very much hope this sourcebook eases the search.

Richard von Götzen

Editor
January 2008

Glossary of terms used

Glossary of terms used

Arrangement fee	A fee charged by the lender to cover its administrative costs. Generally the more complicated the loan, the higher the fee.
Amortization	Capital repayment of the loan. Usually a percentage per annum on top of the loan.
Balloon payment	The largest repayment of a loan usually repaid by instalments. This is normally paid at the beginning and/or end of a loan.
Base rate	The basic rate of interest.
Basis point	One-hundredth of one percent.
Bridging finance	An interim loan enabling the borrower to later refinance the proposal.
Bricks and mortar	The value of the standing asset vacant.
Bullet	Lump sum repayment of the loan at the end of the loan term.
Building society cost of funds	The cost at which the building societies borrow money.
Cap	An arrangement ensuring an upper limit on the interest rate payable on a variable or floating rate loan. The cap comes into play if interest rates move above the level of the cap.
Carried interest	A profit share over and above the return received by a lender or principal once their loan has been repaid.
Collar	Same as a cap only that in addition to an upper limit there is also a lower limit, in which case the borrower continues to pay at collar level — even if interest rates fall.
Cost of funds	The cost of borrowing in the money market.
Covenant	The tenant of the property.
Debt finance	Usually a first charge as mortgage.
Equity finance	Participation by the lender or investor in the project or property.
FHBR	Finance House Base Rate.
Fixed	The rate of interest is fixed at an amount for a specified period.

Floating rate	Interest rate that fluctuates to changes in Libor.
Interest only	A loan without amortization where interest only is payable.
Libor	London Interbank offered rate. The rate at which banks lend to one another.
LTV	Loan to Value. The amount of loan as a percentage of the value.
Margin	The percentage over whatever benchmark at which the lender is prepared to lend money.
Mezzanine finance	A loan over and above the LTV. Because of the risk factor involved in such loans they tend to be expensive.
Structured finance	Bespoke financing.
Pre-let	A binding commitment from an occupier before construction of a building is completed to enter into a lease on specific terms for the building.
Swap	The cost of swapping funds with another lender in the international market. This can be applied to the rate, type and currency etc.

Lender Indices

By Specialisms

The following lenders have indicated they will consider loans in specific markets.

Agriculture

Agricultural Mortgage Corporation Plc
Bank of Scotland (Centrebank)
Barclays Bank Plc
Lloyds TSB
Midland Bank
National Counties Building Society
National Westminster Bank Plc
Northern Bank Ltd
Scottish Agricultural Securities Corporation Plc
The Royal Bank of Scotland
Yorkshire Bank Plc

Investment Properties

Aareal Bank
Abbey National
ABC International Bank Plc
Agricultural Mortgage Corporation Plc
Ahli United Bank (UK) Plc
Alliance & Leicester
Allied Irish Bank
Anglo Irish Bank
Arbuthnot Latham & Co
Ardent Finance
Babylon Business Finance
Bank Leumi (UK) Plc
Bank of East Asia
Bank of Ireland
Bank of Scotland (Centrebank)
Bank of Tokyo — Mitsubishi
Bankers Trust
Barclays Bank Plc
Barclays Property Team
Barnsley Building Society
Bath Building Society
Bear Stearns
Berlin Hyp
BHF-Bank AG
Birmingham Midshires
Bradford & Bingley Plc
Bristol & West Property Finance
Britannia Commercial Lending
British Arab Commercial Bank Ltd
Buckinghamshire Building Society

Butterfield Private Bank
Cambridge Building Society
Canada Life Ltd
Charterhouse Property Finance
Chase Manhattan, JPMorgan
Chesham Building Society
Cheshire Building Society
Citibank International Plc
City Group
Close Property Finance Ltd
Clydesdale Bank
Co-operative Bank
Commercial Acceptances
Coutts & Co
Coventry Building Society
Crédit Agricole Indosuez
Credit Suisse
Cumberland Building Society
Darlington Building Society
Depfa Bank AG
Derbyshire Building Society
Deutsche Bank AG
Deutsche Genossenschafts Hypothekenbank AG
Deutsche Hypo
Deutsche Postbank AG
Dudley Building Society
Dunbar Bank
Dunfermline Building Society
DZ Bank
Eagle Star Mortgages Ltd
Early Shilton Building Society
Ecology Building Society
Egg Mortgages
Erste Bank
Eurohypo
First National Bank
First National Building Society
Furness Building Society
GE Real Estate
GMAC-RFC
Goldman Sachs Real Estate
Hampshire Trust Plc
Henry Ansbacher & Co
Hinckley & Rugby Building Society
Holborn Securities Plc
HSBC Bank Plc
HSH Nordbank AG
Hypo Real Estate Bank International
Hypothekenbank Essen AG
Investec Bank (UK) Ltd
Ipswich Building Society

Irish Permanent International
JPMorgan & Co
Julian Hodge Bank Ltd
Julius Baer International Ltd
KBC Bank NV
Kensington Mortgages
Kent Reliance Building Society
Landesbank Hessen-Thüringen (Helaba)
Leeds Building Society
Leek United Building Society
Legal & General Bank
Lehman Brothers
Liverpool & Victoria
Lloyds TSB
London Mortgage Company
Manchester Building Society
Mansfield Building Society
Market Harborough Building Society
Marsden Building Society
Melton Mowbray Building Society
Merrill Lynch International Ltd
Monmouthshire Building Society
Morgan Guaranty Trust Co (JPMorgan)
Morgan Stanley Dean Witter
Mortgage Trust Ltd
Münchener Hypothekenbank eG
N M Rothschild & Sons
National Counties Building Society
Nationwide Building Society
Natwest — National Westminster Bank Plc
Newbury Building Society
Newcastle Building Society
Nomura Bank International
Nordea
Northern Bank Ltd
Northern Rock Plc
Norwich & Peterborough Building Society
Norwich Union
Nottingham Building Society
Paragon Mortgages Ltd
Principality Building Society
Progressive Building Society
Prudential (Portfolio Managers Ltd)
Saffron Walden Herts & Essex Building Society
Singer & Friedlander Ltd
Skipton Building Society
Société Générale
Stround & Swindon
Sumitomo Bank
Svenska Handelsbanken
Swansea Building Society

The Beverly Building Society
The Chorley Building Society
The Hanley Economic Building Society
The Mortgage Works
The Royal Bank of Scotland
UBS
Ulster Bank Ltd
Union Bank of Switzerland
United Trust Bank Ltd
Vernon Building Socitey
W M Mann & Co (Investments) Ltd
West Bromwich Building Society
Westdeutsche Immobilien Bank
Yorkshire Bank Plc

Refinance
Aareal Bank
Abbey National
ABC International Bank Plc
Agricultural Mortgage Corporation Plc
Ahli United Bank (UK) Plc
Alliance & Leicester
Allied Irish Bank
Anglo Irish Bank
Arbuthnot Latham & Co
Ardent Finance
Babylon Business Finance
Baltic Property Finance Plc
Bank Leumi (UK) Plc
Bank of East Asia
Bank of Ireland
Bank of Scotland (Centrebank)
Bank of Tokyo — Mitsubishi
Bankers Trust
Barclays Bank Plc
Barclays Property Team
Barnsley Building Society
Bath Building Society
Bear Stearns
Berlin Hyp
BHF-Bank AG
Birmingham Midshires
Bradford & Bingley Plc
Bristol & West Property Finance
Britannia Commercial Lending
British Arab Commercial Bank Ltd
Buckinghamshire Building Society
Butterfield Private Bank
Cambridge Building Society
Canada Life Ltd
Charterhouse Property Finance

Chase Manhattan, JPMorgan
Chesham Building Society
Cheshire Building Society
Citibank International Plc
City Group
Close Property Finance Ltd
Clydesdale Bank
Co-operative Bank
Commercial Acceptances
Coutts & Co
Crédit Agricole Indosuez
Credit Suisse
Cumberland Building Society
Darlington Building Society
Depfa Bank AG
Derbyshire Building Society
Deutsche Bank AG
Deutsche Genossenschafts Hypothekenbank AG
Deutsche Hypo
Deutsche Postbank AG
Dudley Building Society
Dunbar Bank
Dunfermline Building Society
DZ Bank
Eagle Star Mortgages Ltd
Early Shilton Building Society
Ecology Building Society
Egg Mortgages
Erste Bank
Eurohypo
First National Bank
First National Building Society
Furness Building Society
GE Real Estate
GMAC-RFC
Goldman Sachs Real Estate
Hampshire Trust Plc
Henry Ansbacher & Co
Heritable Bank Ltd
Hinckley & Rugby Building Society
Holborn Securities Plc
Holmesdale Building Society
HSBC Bank Plc
HSH Nordbank AG
Hypo Real Estate Bank International
Hypothekenbank Essen AG
Investec Bank (UK) Ltd
Ipswich Building Society
Irish Permanent International
JPMorgan & Co
Julian Hodge Bank Ltd

Julius Baer International Ltd
KBC Bank NV
Kensington Mortgages
Kent Reliance Building Society
Landesbank Hessen-Thüringen (Helaba)
Leeds Building Society
Leek United Building Society
Legal & General Bank
Lehman Brothers
Liverpool & Victoria
Lloyds TSB
London Mortgage Company
Manchester Building Society
Mansfield Building Society
Market Harborough Building Society
Marsden Building Society
Melton Mowbray Building Society
Merrill Lynch International Ltd
Monmouthshire Building Society
Morgan Guaranty Trust Co (JPMorgan)
Morgan Stanley Dean Witter
Mortgage Trust Ltd
Münchener Hypothekenbank eG
N M Rothschild & Sons
National Counties Building Society
Nationwide Building Society
Natwest — National Westminster Bank Plc
Newbury Building Society
Newcastle Building Society
Nomura Bank International
Nordea
Northern Bank Ltd
Northern Rock Plc
Norwich & Peterborough Building Society
Norwich Union
Nottingham Building Society
Paragon Mortgages Ltd
Principality Building Society
Progressive Building Society
Prudential (Portfolio Managers Ltd)
Regentsmead Ltd
Saffron Walden Herts & Essex Building Society
Scottish Agricultural Securities Corporation Plc
Singer & Friedlander Ltd
Skipton Building Society
Société Générale
Stround & Swindon
Sumitomo Bank
Svenska Handelsbanken
Swansea Building Society
The Beverly Building Society

The Chorley Building Society
The Hanley Economic Building Society
The Mortgage Works
The Royal Bank of Scotland
UBS
Ulster Bank Ltd
Union Bank of Switzerland
United Trust Bank Ltd
Vernon Building Socitey
W M Mann & Co (Investments) Ltd
West Bromwich Building Society
Westdeutsche Immobilien Bank
Yorkshire Bank Plc

Equity Finance
Aareal Bank
Bank Leumi (UK) Plc
Bank of Ireland
Bankers Trust Company
Bear Stearns
BHF-Bank AG
Charterhouse Property Finance
Chase Manhattan, JPMorgan
City Group
Close Property Finance Ltd
Deutsche Bank AG
GE Real Estate
Goldman Sachs Real Estate
Granville Bank Ltd
Henry Ansbacher & Co
Holborn Securities Plc
HSBC Bank Plc
Intermediate Capital Group Plc
JPMorgan & Co
Lehman Brothers
Morgan Stanley Dean Witter

Owner Occupier
Abbey National
ABC International Bank Plc
Agricultural Mortgage Corporation Plc
Alliance & Leicester
Allied Irish Bank
Anglo Irish Bank
Ardent Finance
Babylon Business Finance
Baltic Property Finance Plc
Bank Leumi (UK) Plc
Bank of East Asia
Bank of Scotland (Centrebank)
Bank of Tokyo — Mitsubishi

Bankers Trust
Barclays Bank Plc
Barclays Property Team
Barnsley Building Society
Bath Building Society
Birmingham Midshires
Bristol & West Property Finance
Britannia Commercial Lending
Buckinghamshire Building Society
Butterfield Private Bank
Cambridge Building Society
Capital Home Loans Ltd
Charterhouse Property Finance
Chelsea Building Society
Chesham Building Society
Cheshire Building Society
Citibank International Plc
City Group
Clydesdale Bank
Co-operative Bank
Commercial Acceptances
Coutts & Co
Crédit Agricole Indosuez
Cumberland Building Society
Darlington Building Society
Depfa Bank AG
Derbyshire Building Society
Deutsche Hypo
Deutsche Postbank AG
Dudley Building Society
Dunfermline Building Society
Eagle Star Mortgages Ltd
Early Shilton Building Society
Ecology Building Society
Egg Mortgages
Eurohypo
First National Bank
First National Building Society
Furness Building Society
Goldman Sachs Real Estate
Hampshire Trust Plc
Hinckley & Rugby Building Society
Holborn Securities Plc
Holmesdale Building Society
HSBC Bank Plc
Hypo Real Estate Bank International
Ipswich Building Society
Irish Permanent International
JPMorgan & Co
Julian Hodge Bank Ltd
Julius Baer International Ltd

KBC Bank NV
Kensington Mortgages
Kent Reliance Building Society
Leeds Building Society
Leek United Building Society
Legal & General Bank
Lehman Brothers
Lloyds TSB
London Mortgage Company
Manchester Building Society
Mansfield Building Society
Market Harborough Building Society
Marsden Building Society
Melton Mowbray Building Society
Monmouthshire Building Society
Morgan Stanley Dean Witter
Mortgage Trust Ltd
National Counties Building Society
Nationwide Building Society
Natwest — National Westminster Bank Plc
Newbury Building Society
Nomura Bank International
Nordea
Northern Bank Ltd
Northern Rock Plc
Norwich & Peterborough Building Society
Nottingham Building Society
Paragon Mortgages Ltd
Principality Building Society
Progressive Building Society
Prudential (Portfolio Managers Ltd)
Saffron Walden Herts & Essex Building Society
Scottish Agricultural Securities Corporation Plc
Singer & Friedlander Ltd
Skipton Building Society
Svenska Handelsbanken
Swansea Building Society
The Beverly Building Society
The Chorley Building Society
The Hanley Economic Building Society
The Mortgage Works
The Royal Bank of Scotland
UBS
Ulster Bank Ltd
Union Bank of Switzerland
United Trust Bank Ltd
Vernon Building Socitey
W M Mann & Co (Investments) Ltd
Yorkshire Bank Plc

Residential Investments

Abbey National
ABC International Bank Plc
Ahli United Bank (UK) Plc
Allied Irish Bank
Anglo Irish Bank
Babylon Business Finance
Bank Leumi (UK) Plc
Bank of East Asia
Bank of Ireland
Barclays Bank Plc
Barclays Property Team
Bath Building Society
Berlin Hyp
Bristol & West Property Finance
Britannia Commercial Lending
Buckinghamshire Building Society
Butterfield Private Bank
Cambridge Building Society
Capital Home Loans Ltd
Charterhouse Property Finance
Chelsea Building Society
Chesham Building Society
Cheshire Building Society
Citibank International Plc
Close Property Finance Ltd
Clydesdale Bank
Commercial Acceptances
Coutts & Co
Coventry Building Society
Cumberland Building Society
Darlington Building Society
Derbyshire Building Society
Deutsche Hypo
Dudley Building Society
Dunbar Bank
Dunfermline Building Society
Early Shilton Building Society
Egg Mortgages
Eurohypo
First National Bank
First National Building Society
Furness Building Society
GE Real Estate
Hampshire Trust Plc
Heritable Bank Ltd
Hinckley & Rugby Building Society
Holborn Securities Plc
Holmesdale Building Society
HSBC Bank Plc
Hypo Real Estate Bank International

Investec Bank (UK) Ltd
Ipswich Building Society
Irish Permanent International
JPMorgan & Co
Julius Baer International Ltd
Kensington Mortgages
Kent Reliance Building Society
Landesbank Hessen-Thüringen (Helaba)
Leeds Building Society
Leek United Building Society
Lloyds TSB
London Mortgage Company
Manchester Building Society
Mansfield Building Society
Market Harborough Building Society
Marsden Building Society
Melton Mowbray Building Society
Merrill Lynch International Ltd
Monmouthshire Building Society
Morgan Stanley Dean Witter
Mortgage Trust Ltd
National Counties Building Society
Nationwide Building Society
Natwest — National Westminster Bank Plc
Newbury Building Society
Newcastle Building Society
Nordea
Northern Bank Ltd
Northern Rock Plc
Nottingham Building Society
Paragon Mortgages Ltd
Principality Building Society
Progressive Building Society
Saffron Walden Herts & Essex Building Society
Sainsbury's Bank
Singer & Friedlander Ltd
Skipton Building Society
Swansea Building Society
The Beverly Building Society
The Chorley Building Society
The Hanley Economic Building Society
The Mortgage Works
The Royal Bank of Scotland
Ulster Bank Ltd
United Trust Bank Ltd
Vernon Building Socitey
W M Mann & Co (Investments) Ltd
West Bromwich Building Society
Westdeutsche Immobilien Bank
Yorkshire Bank Plc

Development Finance
Aareal Bank
ABC International Bank Plc
Agricultural Mortgage Corporation Plc
Ahli United Bank (UK) Plc
Alliance & Leicester
Allied Irish Bank
Anglo Irish Bank
Arbuthnot Latham & Co
Babylon Business Finance
Baltic Property Finance Plc
Bank Leumi (UK) Plc
Bank of Ireland
Bank of Tokyo — Mitsubishi
Barclays Bank Plc
Barclays Property Team
Bath Building Society
Bear Stearns
Berlin Hyp
BHF-Bank AG
Birmingham Midshires
Bradford & Bingley Plc
Bristol & West Property Finance
Butterfield Private Bank
Cambridge Building Society
Charterhouse Property Finance
City Group
Close Property Finance Ltd
Clydesdale Bank
Commercial Acceptances
Coutts & Co
Credit Suisse
Depfa Bank AG
Deutsche Bank AG
Deutsche Genossenschafts Hypothekenbank AG
Deutsche Hypo
Deutsche Postbank AG
Dunbar Bank
Dunfermline Building Society
DZ Bank
Egg Mortgages
Erste Bank
Eurohypo
Furness Building Society
GE Real Estate
GMAC-RFC
Goldman Sachs Real Estate
Hampshire Trust Plc
Henry Ansbacher & Co
Heritable Bank Ltd
Holborn Securities Plc

HSBC Bank Plc
Hypo Real Estate Bank International
Hypothekenbank Essen AG
Intermediate Capital Group Plc
Investec Bank (UK) Ltd
JPMorgan & Co
Julian Hodge Bank Ltd
KBC Bank NV
Kent Reliance Building Society
Landesbank Hessen-Th¸ringen (Helaba)
Lehman Brothers
Lloyds TSB
London Mortgage Company
Merrill Lynch International Ltd
Mortgage Trust Ltd
Münchener Hypothekenbank eG
N M Rothschild & Sons
Natwest — National Westminster Bank Plc
Nordea
Northern Bank Ltd
Northern Rock Plc
Paragon Mortgages Ltd
Principality Building Society
Regentsmead Ltd
Saffron Walden Herts & Essex Building Society
Singer & Friedlander Ltd
Société Générale
Sumitomo Bank
The Mortgage Works
The Royal Bank of Scotland
Ulster Bank Ltd
Union Bank of Switzerland
United Trust Bank Ltd
Vernon Building Socitey
W M Mann & Co (Investments) Ltd
West Bromwich Building Society
Westdeutsche Immobilien Bank
Yorkshire Bank Plc

Residential Development
ABC International Bank Plc
Ahli United Bank (UK) Plc
Allied Irish Bank
Anglo Irish Bank
Arbuthnot Latham & Co
Babylon Business Finance
Bank Leumi (UK) Plc
Bank of East Asia
Bank of Ireland
Barclays Bank Plc
Barclays Property Team

Bath Building Society
BHF-Bank AG
Bradford & Bingley Plc
Bristol & West Property Finance
Butterfield Private Bank
Cambridge Building Society
Canada Life Ltd
Charterhouse Property Finance
Chase Manhattan, JPMorgan
City Group
Close Property Finance Ltd
Clydesdale Bank
Commercial Acceptances
Coutts & Co
Deutsche Hypo
Dunbar Bank
Dunfermline Building Society
Egg Mortgages
Eurohypo
First National Building Society
GMAC-RFC
Hampshire Trust Plc
Heritable Bank Ltd
Holborn Securities Plc
HSBC Bank Plc
HSH Nordbank AG
Hypo Real Estate Bank International
Hypothekenbank Essen AG
Investec Bank (UK) Ltd
JPMorgan & Co
Julian Hodge Bank Ltd
Julius Baer International Ltd
Kent Reliance Building Society
Landesbank Hessen-Thüringen (Helaba)
Leek United Building Society
Lehman Brothers
London Mortgage Company
Manchester Building Society
Mansfield Building Society
Merrill Lynch International Ltd
Mortgage Trust Ltd
Münchener Hypothekenbank eG
Natwest — National Westminster Bank Plc
Northern Bank Ltd
Paragon Mortgages Ltd
Principality Building Society
Regentsmead Ltd
Saffron Walden Herts & Essex Building Society
Sainsbury's Bank
Singer & Friedlander Ltd
The Mortgage Works

The Royal Bank of Scotland
Ulster Bank Ltd
United Trust Bank Ltd
W M Mann & Co (Investments) Ltd
Westdeutsche Immobilien Bank
Yorkshire Bank Plc

Hotels and Leisure
Aareal Bank
ABC International Bank Plc
Ahli United Bank (UK) Plc
Allied Irish Bank
Anglo Irish Bank
Ardent Finance
Babylon Business Finance
Bank Leumi (UK) Plc
Bank of East Asia
Bank of Ireland
Bank of Scotland (Centrebank)
Barclays Bank Plc
Barclays Property Team
Bath Building Society
Bear Stearns
BHF-Bank AG
Birmingham Midshires
Bristol & West Property Finance
Britannia Commercial Lending
Cambridge Building Society
Canada Life Ltd
Charterhouse Property Finance
Chase Manhattan, JPMorgan
Chesham Building Society
Cheshire Building Society
Citibank International Plc
City Group
Clydesdale Bank
Co-operative Bank
Credit Suisse
Darlington Building Society
Depfa Bank AG
Deutsche Genossenschafts Hypothekenbank AG
Deutsche Postbank AG
Dunfermline Building Society
DZ Bank
Egg Mortgages
Eurohypo
First National Bank
First National Building Society
Furness Building Society
GE Real Estate
GMAC-RFC

Goldman Sachs Real Estate
Hampshire Trust Plc
Holborn Securities Plc
HSBC Bank Plc
HSH Nordbank AG
Hypo Real Estate Bank International
Hypothekenbank Essen AG
Intermediate Capital Group Plc
JPMorgan & Co
Julius Baer International Ltd
Kent Reliance Building Society
Lehman Brothers
London Mortgage Company
Manchester Building Society
Mansfield Building Society
Marsden Building Society
Merrill Lynch International Ltd
Morgan Guaranty Trust Co (J P Morgan)
Morgan Stanley Dean Witter
Mortgage Trust Ltd
Münchener Hypothekenbank eG
National Counties Building Society
Natwest — National Westminster Bank Plc
Newbury Building Society
Newcastle Building Society
Nordea
Northern Bank Ltd
Northern Rock Plc
Norwich & Peterborough Building Society
Nottingham Building Society
Paragon Mortgages Ltd
Prudential (Portfolio Managers Ltd)
Saffron Walden Herts & Essex Building Society
Singer & Friedlander Ltd
Skipton Building Society
Société Générale
Swansea Building Society
The Mortgage Works
The Royal Bank of Scotland
Ulster Bank Ltd
Union Bank of Switzerland
W M Mann & Co (Investments) Ltd
West Bromwich Building Society
Westdeutsche Immobilien Bank
Yorkshire Bank Plc

Healthcare

ABC International Bank Plc
Ahli United Bank (UK) Plc
Allied Irish Bank
Anglo Irish Bank

Babylon Business Finance
Bank of East Asia
Bank of Ireland
Bank of Scotland (Centrebank)
Barclays Bank Plc
Bath Building Society
Bear Stearns
BHF-Bank AG
Birmingham Midshires
Bristol & West Property Finance
Britannia Commercial Lending
Buckinghamshire Building Society
Cambridge Building Society
Chase Manhattan, JPMorgan
Chesham Building Society
Cheshire Building Society
Citibank International Plc
Clydesdale Bank
Co-operative Bank
Coutts & Co
Credit Suisse
Darlington Building Society
Derbyshire Building Society
Deutsche Postbank AG
Dunfermline Building Society
DZ Bank
Egg Mortgages
Eurohypo
First National Building Society
GE Real Estate
GMAC-RFC
Hampshire Trust Plc
Holborn Securities Plc
HSH Nordbank AG
Hypo Real Estate Bank International
Hypothekenbank Essen AG
Intermediate Capital Group Plc
JPMorgan & Co
Julian Hodge Bank Ltd
Kensington Mortgages
Kent Reliance Building Society
Lehman Brothers
Lloyds TSB
London Mortgage Company
Manchester Building Society
Merrill Lynch International Ltd
Morgan Guaranty Trust Co (JPMorgan)
Mortgage Trust Ltd
Münchener Hypothekenbank eG
National Counties Building Society
Natwest — National Westminster Bank Plc

Newbury Building Society
Newcastle Building Society
Northern Bank Ltd
Northern Rock Plc
Norwich & Peterborough Building Society
Nottingham Building Society
Paragon Mortgages Ltd
Prudential (Portfolio Managers Ltd)
Singer & Friedlander Ltd
Skipton Building Society
The Mortgage Works
The Royal Bank of Scotland
Ulster Bank Ltd
United Trust Bank Ltd
W M Mann & Co (Investments) Ltd
West Bromwich Building Society
Westdeutsche Immobilien Bank
Yorkshire Bank Plc

Providers of Bridging Loans
ABC International Bank Plc
Ahli United Bank (UK) Plc
Allied Irish Bank
Anglo Irish Bank
Baltic Property Finance Plc
Bank Leumi (UK) Plc
Bank of Ireland
Bank of Tokyo — Mitsubishi
Bankers Trust
Barclays Bank Plc
Barclays Property Team
Bear Stearns
BHF-Bank AG
Bristol & West Property Finance
Charterhouse Property Finance
Chase Manhattan, JPMorgan
City Group
Commercial Acceptances
Coutts & Co
Credit Suisse
Deutsche Bank AG
Deutsche Postbank AG
DZ Bank
Egg Mortgages
Eurohypo
GMAC-RFC
Goldman Sachs Real Estate
Hampshire Trust Plc
Henry Ansbacher & Co
Heritable Bank Ltd
Holborn Securities Plc

HSBC Bank Plc
Hypo Real Estate Bank International
Hypothekenbank Essen AG
Intermediate Capital Group Plc
Investec Bank (UK) Ltd
JPMorgan & Co
Julian Hodge Bank Ltd
Julius Baer International Ltd
Kensington Mortgages
Lehman Brothers
London Mortgage Company
Manchester Building Society
Merrill Lynch International Ltd
Morgan Guaranty Trust Co (JPMorgan)
Münchener Hypothekenbank eG
N M Rothschild & Sons
Natwest — National Westminster Bank Plc
Nomura Bank International
Paragon Mortgages Ltd
Regentsmead Ltd
Singer & Friedlander Ltd
The Mortgage Works
Ulster Bank Ltd
Union Bank of Switzerland
United Trust Bank Ltd
W M Mann & Co (Investments) Ltd

Providers of Mezzanine Finance

ABC International Bank Plc
Ahli United Bank (UK) Plc
Allied Irish Bank
Anglo Irish Bank
Baltic Property Finance Plc
Bank Leumi (UK) Plc
Bank of Ireland
Bank of Tokyo — Mitsubishi
Bankers Trust
Barclays Bank Plc
Barclays Property Team
Bear Stearns
BHF-Bank AG
Bristol & West Property Finance
Britannia Commercial Lending
Charterhouse Property Finance
Chase Manhattan, JPMorgan
City Group
Close Property Finance Ltd
Commercial Acceptances
Coutts & Co
Credit Suisse
Deutsche Bank AG

Deutsche Postbank AG
DZ Bank
Egg Mortgages
Erste Bank
Eurohypo
GE Real Estate
GMAC-RFC
Goldman Sachs Real Estate
Hampshire Trust Plc
Henry Ansbacher & Co
Heritable Bank Ltd
Holborn Securities Plc
HSBC Bank Plc
Hypo Real Estate Bank International
Hypothekenbank Essen AG
Intermediate Capital Group Plc
Investec Bank (UK) Ltd
JPMorgan & Co
Julian Hodge Bank Ltd
Julius Baer International Ltd
Kensington Mortgages
Lehman Brothers
London Mortgage Company
Manchester Building Society
Merrill Lynch International Ltd
Morgan Guaranty Trust Co (JPMorgan)
Münchener Hypothekenbank eG
N M Rothschild & Sons
Natwest — National Westminster Bank Plc
Nomura Bank International
Paragon Mortgages Ltd
Singer & Friedlander Ltd
The Mortgage Works
Ulster Bank Ltd
Union Bank of Switzerland
United Trust Bank Ltd
W M Mann & Co (Investments) Ltd

Speculative or Part Pre-let Development
Commercial Acceptances
Deutsche Bank AG
Deutsche Genossenschafts Hypothekenbank AG
Deutsche Hypo
Erste Bank
Furness Building Society
Landesbank Hessen-Thüringen (Helaba)
Lloyds TSB
Merrill Lynch International Ltd
Nordea
Northern Rock Plc
The Royal Bank of Scotland

Sumitomo Bank
Union Bank of Switzerland
Vernon Building Society

Regional Lenders

Bath Building Society	30-mile radius of Bath
Cambridge Building Society	50-mile radius of Cambridge
Coventry Building Society	50-mile radius of Coventry
Darlington Building Society	South Durham and North Yorkshire
Dunfermline Building Society	North England and Scotland
First National Building Society	Irish Republic
Furness Building Society	100-mile radius of Barrow-in-Furness
Hinckley & Rugby Building Society	East Midlands
Leek United Building Society	60-mile radius of Leek
Manchester Building Society	50-mile radius of Manchester
Northern Bank Ltd	North Ireland only
Marsden Building Society	100-mile radius of Marsden
Newbury Building Society	60-mile radius of Newbury
Principality Building Society	Wales only
Ulster Bank Ltd	North Ireland and Republic of Ireland only
W M Mann & Co (Investments) Ltd	Scotland only

Retail Properties

Aareal Bank
Abbey National Plc
Ahli United Bank (UK) Plc
Alliance & Leicester (Giro Bank)
Allied Irish Bank
Anglo Irish Bank Corporation Plc
Arbuthnot Latham & Co
BHF-Bank AG
Baltic Property Finance Plc
Bank Julius Baer & Co
Bank of Scotland (Centrebank)
Bank of Tokyo — Mitsubishi
Bankers Trust Company
Barclays Bank Plc
Barnsley Building Society
Bath Building Society
Birmingham Midshires
Bradford & Bingley Plc
Bristol & West Plc
British Arab Commercial Bank Ltd
Brown Shipley & Co
Cambridge Building Society
Canada Life Ltd
Charterhouse Property Finance Ltd
Chase Manhattan, JPMorgan
Cheshire Building Society
Citibank International Plc

City Group
Close Property Finance Ltd
Clydesdale Bank
Commercial Acceptances
Co-operative Bank
Coventry Building Society
Crédit Agricole Indosuez
Darlington Building Society
Depfa Bank AG
Derbyshire Building Society
Deutsche Bank AG
Deutsche Hypo
Dunbar Group Plc
Dunfermline Building Society
Eagle Star Mortgages Ltd
Erste Bank
Eurohypo
First National Bank
First National Building Society
Furness Building Society
GE Real Estate
Goldman Sachs Real Estate
Hamburgische Landesbank
Hampshire Trust Plc
Henry Ansbacher & Co
Hinckley & Rugby Building Society
Julian Hodge Bank Ltd
Holborn Securities Plc
HSBC Bank Plc
Investec Bank (UK) Ltd
JPMorgan & Co
Julian Hodge Bank Ltd
KBC Bank NV
Kent Reliance Building Society
Landesbank Hessen Thuringen (Helaba)
Leek United Building Society
Legal & General Bank
Leopold Joseph & Sons Ltd
Liverpool & Victoria
Lloyds TSB
Manchester Building Society
Mansfield Building Society
Marsden Building Society
Merrill Lynch International Ltd
Morgan Stanley Dean Witter
NM Rothschild & Sons
National Counties Building Society
National Westminster Bank Plc
Nationwide Building Society
Newbury Building Society
Newcastle Building Society

Nomura Bank International
Nordea
Northern Bank Ltd
Northern Rock Plc
Norwich & Peterborough Building Society
Norwich Union
Principality Building Society
Prudential (Portfolio Managers Ltd)
Regentsmead Ltd
Saffron Walden Herts & Essex Building Society
Société Générale
Sumitomo Bank
Svenska Handelsbanken
The Bank of East Asia
The Royal Bank of Scotland
UBS
Ulster Bank Ltd
Vernon Building Society
WM Mann & Co (Investments) Ltd
West Bromwich Building Society
Westdeutsche Immobilien Bank
Wintrust Securities Ltd
Yorkshire Bank Plc

By Maximum Loan To Value

These lenders stipulate a ceiling to their lending as a proportion of the value of the property.

100%

Babylon Business Finance
Baltic Property Finance Plc
Bankers Trust
Bear Stearns
Capital Home Loans Ltd
Charterhouse Property Finance
Chase Manhattan, JPMorgan
Citibank International Plc
City Group
Depfa Bank AG
Deutsche Bank AG
Egg Mortgages
Goldman Sachs Real Estate
Hypo Real Estate Bank International
Intermediate Capital Group Plc
JPMorgan & Co
Lehman Brothers
Merrill Lynch International Ltd
Nomura Bank International
UBS

95%
Allied Irish Bank
Ardent Finance
Barnsley Building Society
Bath Building Society
Chelsea Building Society
Erste Bank
Henry Ansbacher & Co
Morgan Stanley Dean Witter

90%
Abbey National
ABC International Bank Plc
Anglo Irish Bank
Bank of Ireland
Barclays Bank Plc
Barclays Property Team
Berlin Hyp
Canada Life Ltd
Coutts & Co
Deutsche Postbank AG
DZ Bank
Eurohypo
First National Bank
GE Real Estate
GMAC-RFC
HSBC Bank Plc
HSH Nordbank AG
Investec Bank (UK) Ltd
Mortgage Trust Ltd
Münchener Hypothekenbank eG
N M Rothschild & Sons
Natwest — National Westminster Bank Plc
Norwich Union
Paragon Mortgages Ltd
Singer & Friedlander Ltd
United Trust Bank Ltd

85%
Aareal Bank
Arbuthnot Latham & Co
Bank Leumi (UK) Plc
Bank of East Asia
Bank of Scotland (Centrebank)
Birmingham Midshires
Bradford & Bingley Plc
Britannia Commercial Lending
Butterfield Private Bank
Chesham Building Society
Coventry Building Society
Credit Suisse

Dudley Building Society
Eagle Star Mortgages Ltd
Heritable Bank Ltd
Holborn Securities Plc
Hypothekenbank Essen AG
Julius Baer International Ltd
Kent Reliance Building Society
Lloyds TSB
London Mortgage Company
Manchester Building Society
National Counties Building Society
Newbury Building Society
Newcastle Building Society
Northern Bank Ltd
Northern Rock Plc
Norwich & Peterborough Building Society
Principality Building Society
Prudential (Portfolio Managers Ltd)
Saffron Walden Herts & Essex Building Society
The Mortgage Works
Ulster Bank Ltd
Vernon Building Socitey
Yorkshire Bank Plc

80%
Ahli United Bank (UK) Plc
Alliance & Leicester
Bank of Tokyo — Mitsubishi
BHF-Bank AG
Buckinghamshire Building Society
Clydesdale Bank
Derbyshire Building Society
Deutsche Genossenschafts Hypothekenbank AG
Dunbar Bank
Early Shilton Building Society
First National Building Society
Holmesdale Building Society
Ipswich Building Society
Irish Permanent International
KBC Bank NV
Kensington Mortgages
Landesbank Hessen-Thüringen (Helaba)
Leeds Building Society
Legal & General Bank
Liverpool & Victoria
Mansfield Building Society
Market Harborough Building Society
Melton Mowbray Building Society
Monmouthshire Building Society
Nationwide Building Society
Nordea

Nottingham Building Society
Progressive Building Society
Regentsmead Ltd
Swansea Building Society
The Hanley Economic Building Society
West Bromwich Building Society
Westdeutsche Immobilien Bank

75%
British Arab Commercial Bank Ltd
Cambridge Building Society
Cheshire Building Society
Close Property Finance Ltd
Co-operative Bank
Crédit Agricole Indosuez
Cumberland Building Society
Darlington Building Society
Deutsche Hypo
Dunfermline Building Society
Ecology Building Society
Furness Building Society
Hampshire Trust Plc
Hinckley & Rugby Building Society
Julian Hodge Bank Ltd
Leek United Building Society
Marsden Building Society
Sainsbury's Bank
Skipton Building Society
Société Générale
Stround & Swindon
Sumitomo Bank
Svenska Handelsbanken
The Beverly Building Society
The Chorley Building Society

70%
Commercial Acceptances
W M Mann & Co (Investments) Ltd

65%
Scottish Agricultural Securities Corporation Plc

50%
Agricultural Mortgage Corporation Plc

By Size of Loan (more than £10m)

These lenders will consider loans of more than £10m.

Aareal Bank
Abbey National
ABC International Bank Plc

Agricultural Mortgage Corporation Plc
Ahli United Bank (UK) Plc
Alliance & Leicester
Allied Irish Bank
Anglo Irish Bank
Arbuthnot Latham & Co
Babylon Business Finance
Baltic Property Finance Plc
Bank Leumi (UK) Plc
Bank of Ireland
Bank of Scotland (Centrebank)
Bank of Tokyo — Mitsubishi
Bankers Trust
Barclays Bank Plc
Barclays Property Team
Bear Stearns
Berlin Hyp
BHF-Bank AG
Birmingham Midshires
Bradford & Bingley Plc
Bristol & West Property Finance
Britannia Commercial Lending
British Arab Commercial Bank Ltd
Canada Life Ltd
Charterhouse Property Finance
Chase Manhattan, JPMorgan
Citibank International Plc
City Group
Close Property Finance Ltd
Clydesdale Bank
Co-operative Bank
Coutts & Co
Crédit Agricole Indosuez
Credit Suisse
Derbyshire Building Society
Deutsche Bank AG
Deutsche Genossenschafts Hypothekenbank AG
Deutsche Hypo
Deutsche Postbank AG
Dunbar Bank
DZ Bank
Egg Mortgages
Erste Bank
Eurohypo
GE Real Estate
GMAC-RFC
Goldman Sachs Real Estate
Heritable Bank Ltd
HSBC Bank Plc
HSH Nordbank AG
Hypo Real Estate Bank International

Hypothekenbank Essen AG
Intermediate Capital Group Plc
Investec Bank (UK) Ltd
Irish Permanent International
JPMorgan & Co
Julian Hodge Bank Ltd
KBC Bank NV
Landesbank Hessen-Thüringen (Helaba)
Legal & General Bank
Lehman Brothers
Lloyds TSB
Merrill Lynch International Ltd
Morgan Guaranty Trust Co (JPMorgan)
Morgan Stanley Dean Witter
Mortgage Trust Ltd
Münchener Hypothekenbank eG
N M Rothschild & Sons
National Counties Building Society
Nationwide Building Society
Natwest — National Westminster Bank Plc
Newcastle Building Society
Nomura Bank International
Nordea
Northern Bank Ltd
Northern Rock Plc
Norwich & Peterborough Building Society
Norwich Union
Paragon Mortgages Ltd
Prudential (Portfolio Managers Ltd)
Singer & Friedlander Ltd
Skipton Building Society
Société Générale
Stround & Swindon
Sumitomo Bank
Svenska Handelsbanken
The Royal Bank of Scotland
UBS
West Bromwich Building Society
Westdeutsche Immobilien Bank
Yorkshire Bank Plc

By Margin Spread
Minimum margin 1.5% and below.

Aareal Bank
Abbey National
Agricultural Mortgage Corporation Plc
Ahli United Bank (UK) Plc
Allied Irish Bank
Anglo Irish Bank
Ardent Finance

Babylon Business Finance
Bank of East Asia
Bank of Ireland
Bank of Scotland (Centrebank)
Bank of Tokyo — Mitsubishi
Barclays Bank Plc
Barclays Property Team
Bath Building Society
Berlin HYP
BHF-Bank AG
Birmingham Midshires
British Arab Commercial Bank Ltd
Buckinghamshire Building Society
Butterfield Private Bank
Charterhouse Property Finance
Chase Manhattan, JPMorgan
Chesham Building Society
Cheshire Building Society
Citibank International Plc
City Group
Clydesdale Bank
Coutts & Co
Crédit Agricole Indosuez
Credit Suisse
Cumberland Building Society
Darlington Building Society
Depfa Bank AG
Derbyshire Building Society
Deutsche Genossenschafts Hypothekenbank AG
Deutsche Hypo
Deutsche Postbank AG
Dudley Building Society
Dunfermline Building Society
Eagle Star Mortgages Ltd
Early Shilton Building Society
Ecology Building Society
Egg Mortgages
Erste Bank
Eurohypo
First National Bank
Furness Building Society
GE Real Estate
Henry Ansbacher & Co
Hinckley & Rugby Building Society
Holborn Securities Plc
Holmesdale Building Society
HSBC Bank Plc
HSH Nordbank AG
Hypo Real Estate Bank International
Hypothekenbank Essen AG
Investec Bank (UK) Ltd

Ipswich Building Society
Irish Permanent International
Julius Baer International Ltd
KBC Bank NV
Kensington Mortgages
Kent Reliance Building Society
Landesbank Hessen-Thüringen (Helaba)
Leeds Building Society
Leek United Building Society
Legal & General Bank
Lehman Brothers
Lloyds TSB
London Mortgage Company
Market Harborough Building Society
Marsden Building Society
Melton Mowbray Building Society
Monmouthshire Building Society
Morgan Stanley Dean Witter
Mortgage Trust Ltd
N M Rothschild & Sons
National Counties Building Society
Newbury Building Society
Newcastle Building Society
Nomura Bank International
Nordea
Northern Bank Ltd
Northern Rock Plc
Norwich & Peterborough Building Society
Nottingham Building Society
Paragon Mortgages Ltd
Progressive Bulding Society
Prudential (Portfolio Managers Ltd)
Regentsmead Ltd
Saffron Walden Herts & Essex Building Society
Sainsbury's Bank
Société Générale
Stroud & Swindon
Svenska Handelsbanken
Swansea Building Society
The Beverly Building Society
The Chorley Building Society
The Hanley Economic Building Society
The Mortgage Works
UBS
Vernon Building Society
West Bromwich Building Society
Westdeutsche Immobilien Bank

COMPANY LISTINGS

Aareal Bank

38 Lombard Street
London
EC3V 9BS

Tel 020 7456 9200
Fax 020 7929 5055
www.aareal-bank.com

Contact
Paul Stone
General Manager
020 7456 9227
paul.stone@aareal-bank.com

Criteria

1	Maximum Loan to Value	85%
2	Benchmark Lending Rate	Libor
3	Margin Spread	50–100bp
4	Fixed	Yes
5	Minimum Loan	£10,000,000
6	Maximum Loan	None
7	Arrangement Fee Spread	N/A
8	Minimum Length of Lease Required	Depends
9	Usual Length of Loan	5–10 years
10	Investment	Yes
11	Refinance	Yes
12	Owner Occupier	No
13	Residential Investment	No
14	Development Finance	Yes
15	Residential Development	No
16	Hotels and Leisure	Yes
17	Healthcare	No
18	Bridging and/or Mezzanine Finance	No

Abbey National

Abbey National House
2 Triton Square
Regent's Place
London
NW1 3AN

Tel 0870 607 6000
www.abbey.com

Contact
Andrew Turzynski
Head of Property Finance
020 8416 7810
andrew.turzynsky@abbey.com

Criteria

1	Maximum Loan to Value	90%
2	Benchmark Lending Rate	Libor
3	Margin Spread	75–125bp
4	Fixed	Yes
5	Minimum Loan	£50,001
6	Maximum Loan	None
7	Arrangement Fee Spread	0.25–0.75%
8	Minimum Length of Lease Required	Varies
9	Usual Length of Loan	3–25 years
10	Investment	Yes
11	Refinance	Yes
12	Owner Occupier	Yes
13	Residential Investment	Yes
14	Development Finance	No
15	Residential Development	No
16	Hotels and Leisure	No
17	Healthcare	No
18	Bridging and/or Mezzanine Finance	No

Remarks

Part of Santander Group.

ABC International Bank Plc

ABC International Bank Plc
Arab Banking Corporation House
1–5 Moorgate
London
EC2R 6AB

Tel 020 7776 4000
Fax 020 7606 9987
www.arabbanking.com

Contact

Keith Leach
European Head of Project
& Structured Finance
020 7776 4028

Peter Dowham
Head of Finance

Criteria

1	Maximum Loan to Value	90%
2	Benchmark Lending Rate	Libor
3	Margin Spread	50–100bp
4	Fixed	Yes
5	Minimum Loan	£250,000
6	Maximum Loan	None
7	Arrangement Fee Spread	0.50–1.0%
8	Minimum Length of Lease Required	50 years
9	Usual Length of Loan	5–10 years
10	Investment	Yes
11	Refinance	Yes
12	Owner Occupier	Yes
13	Residential Investment	Yes
14	Development Finance	Yes
15	Residential Development	Yes
16	Hotels and Leisure	Yes
17	Healthcare	Yes
18	Bridging and/or Mezzanine Finance	Yes

Agricultural Mortgage Corporation Plc

A M C House
Chantry Street
Andover
Hampshire
SP10 1DE

Tel 01264 334344
01264 334747
Fax 01264 334614
www.amconline.co.uk
info@amcplc.com

Contact
Sarah Clifford
Business Manager

Amanda Sheppard
Business Manager

Ian Borrett

Lorraine Lawrie

Criteria

1	Maximum Loan to Value	50%
2	Benchmark Lending Rate	Base + 2%
3	Margin Spread	100–125bp
4	Fixed	Yes
5	Minimum Loan	£50,000
6	Maximum Loan	None
7	Arrangement Fee Spread	0.5–1.0%
8	Minimum Length of Lease Required	30 years
9	Usual Length of Loan	5–40 years
10	Investment	Yes
11	Refinance	Yes
12	Owner Occupier	Yes
13	Residential Investment	No
14	Development Finance	Yes
15	Residential Development	No
16	Hotels and Leisure	No
17	Healthcare	No
18	Bridging and/or Mezzanine Finance	No

Remarks

Lends to commercial farming and rural business, for any commercial purpose.

Ahli United Bank (UK) Plc

35 Portman Square
London
W1H 6LR

Tel 020 7487 6500
Fax 020 7487 6593
www.ahliunited.com
info@ahliunited.com

Contact
Rupert Kirby
020 7487 6992
rupert.kirby@ahliunited.com

Rob Morgan
020 7487 6482
rob.morgan@ahliunited.com

Patrick Ormond
020 7487 6593
patrick.ormond@ahliunited.com

Josh Greig
020 7487 6470
josh.greig@ahliunited.com

Criteria

1	Maximum Loan to Value	80%
2	Benchmark Lending Rate	Libor
3	Margin Spread	1.25–2.50%
4	Fixed	Yes
5	Minimum Loan	£2,500,000
6	Maximum Loan	£25,000,000
7	Arrangement Fee Spread	0.50–1.0%
8	Minimum Length of Lease Required	Negotiable
9	Usual Length of Loan	1–10 years
10	Investment	Possibly
11	Refinance	No
12	Owner Occupier	Yes
13	Residential Investment	Yes
14	Development Finance	Yes
15	Residential Development	Yes
16	Hotels and Leisure	Yes
17	Healthcare	Yes
18	Bridging and/or Mezzanine Finance	Yes

Alliance & Leicester

Carlton Park
Narborough
Leicester
LE19 0AL

Fax 0121 237 6150
www.alliance-leicester.co.uk

Contact

Stuart Wilson
Head of Property Finance
0121 237 6282
stuart.wilson@alliance-leicester.co.

Paul James
Relationship Manager
0121 237 6270

Criteria

1	Maximum Loan to Value	80%
2	Benchmark Lending Rate	Base
3	Margin Spread	2–3%
4	Fixed	Yes
5	Minimum Loan	£25,000
6	Maximum Loan	None
7	Arrangement Fee Spread	1%
8	Minimum Length of Lease Required	N/A
9	Usual Length of Loan	1–10 years
10	Investment	Yes
11	Refinance	Yes
12	Owner Occupier	Yes
13	Residential Investment	No
14	Development Finance	Yes
15	Residential Development	No
16	Hotels and Leisure	No
17	Healthcare	No
18	Bridging and/or Mezzanine Finance	No

Allied Irish Bank

9/10 Angel Court
London
EC2R 7AB

Tel 020 7606 4900
Fax 020 7606 4966
www.aibgroup.com

Contact
Robin Ingram
Chief of City Office

Criteria

1	Maximum Loan to Value	95%
2	Benchmark Lending Rate	Libor
3	Margin Spread	1.0–1.50%
4	Fixed	Yes
5	Minimum Loan	£125,000
6	Maximum Loan	None
7	Arrangement Fee Spread	1%
8	Minimum Length of Lease Required	Depends
9	Usual Length of Loan	5–15 years
10	Investment	Yes
11	Refinance	Yes
12	Owner Occupier	Yes
13	Residential Investment	Yes
14	Development Finance	Yes
15	Residential Development	Yes
16	Hotels and Leisure	Yes
17	Healthcare	Yes
18	Bridging and/or Mezzanine Finance	Yes

Anglo Irish Bank

10 Old Jewry
London
EC2R 8DN

Tel 020 7710 7000
Fax 020 7710 7050
www.angloirishbank.co.uk

Contact

Peter Ralph
Director of UK Lending
peter.ralph@angloirishbank.co.uk

Dean Harris
Property Finance
020 7710 7174

Mark Cotter
Property Finance
020 7710 7008

Criteria

1	Maximum Loan to Value	90%
2	Benchmark Lending Rate	Libor
3	Margin Spread	0.5–2.0%
4	Fixed	Yes
5	Minimum Loan	£500,000
6	Maximum Loan	None
7	Arrangement Fee Spread	0.2–0.5%
8	Minimum Length of Lease Required	Depends
9	Usual Length of Loan	3–10 years
10	Investment	Yes
11	Refinance	Yes
12	Owner Occupier	Yes
13	Residential Investment	Yes
14	Development Finance	Yes
15	Residential Development	Yes
16	Hotels and Leisure	Yes
17	Healthcare	Yes
18	Bridging and/or Mezzanine Finance	Yes

Henry Ansbacher & Co

Two London Bridge
London
SE1 9RA

Tel 020 7089 4700
Fax 020 7089 4850
www.ansbacher.com

Contact
Robert Morral
Head of Specialist Lending
020 7089 4700
robert.morrall@ansbacher.com

Criteria

1	Maximum Loan to Value	95%
2	Benchmark Lending Rate	3 month Libor
3	Margin Spread	1.5–2.0%
4	Fixed	Yes
5	Minimum Loan	£500,000
6	Maximum Loan	£5,000,000
7	Arrangement Fee Spread	0.5–1.0%
8	Minimum Length of Lease Required	10 years
9	Usual Length of Loan	6 months–5 years
10	Investment	Yes
11	Refinance	Yes
12	Owner Occupier	No
13	Residential Investment	No
14	Development Finance	Yes
15	Residential Development	No
16	Hotels and Leisure	No
17	Healthcare	No
18	Bridging and/or Mezzanine Finance	Yes

Remarks

Mezzanine lending preferred or equity with profits share.

Arbuthnot Latham & Co

Arbuthnot House
20 Ropemaker Street
London
EC2Y 9AR

Tel 020 7012 2500
Fax 020 7012 2501
www.arbuthnotlatham.co.uk

Contact
Martin Calder
Head of Property Finance
020 7012 2522

Criteria

1	Maximum Loan to Value	85%
2	Benchmark Lending Rate	Base
3	Margin Spread	2–3%
4	Fixed	No
5	Minimum Loan	£125,000
6	Maximum Loan	£10,000,000
7	Arrangement Fee Spread	1%
8	Minimum Length of Lease Required	30 years remaining
9	Usual Length of Loan	1–5 years
10	Investment	Yes
11	Refinance	Yes
12	Owner Occupier	No
13	Residential Investment	No
14	Development Finance	Yes
15	Residential Development	Yes
16	Hotels and Leisure	No
17	Healthcare	No
18	Bridging and/or Mezzanine Finance	No

Ardent Finance

Ardent House
Home Farm, Mill Lane
Stillington,
York
YO61 1NG

Tel 0134 7811547
Fax 0134 7811425
www.ardentfinance.com

Contact
Chris Tordoff
Mortgage Specialist
christo@ardentfinance.com

Criteria

1	Maximum Loan to Value	95%
2	Benchmark Lending Rate	Libor
3	Margin Spread	0.5–1.0%
4	Fixed	No
5	Minimum Loan	£100,000
6	Maximum Loan	£5,000,000
7	Arrangement Fee Spread	1%
8	Minimum Length of Lease Required	Varies
9	Usual Length of Loan	3–10 years
10	Investment	Yes
11	Refinance	Yes
12	Owner Occupier	Yes
13	Residential Investment	No
14	Development Finance	No
15	Residential Development	No
16	Hotels and Leisure	Yes
17	Healthcare	No
18	Bridging and/or Mezzanine Finance	No

Babylon Business Finance

1 Berkeley Street
London
W1J 8DJ

Tel 020 7016 9775
Fax 020 7016 9100
www.babylonbusinessfinance.com

Contact
Joel Inbakumar
Head of Business Finance Division
joel@babylonbusinessfinance.com

Criteria

1	Maximum Loan to Value	100%
2	Benchmark Lending Rate	Libor
3	Margin Spread	0.25–1.0%
4	Fixed	Yes
5	Minimum Loan	£25,000
6	Maximum Loan	None
7	Arrangement Fee Spread	0.50–1.0%
8	Minimum Length of Lease Required	Varies
9	Usual Length of Loan	1–25 years
10	Investment	Yes
11	Refinance	Yes
12	Owner Occupier	Yes
13	Residential Investment	Yes
14	Development Finance	Yes
15	Residential Development	Yes
16	Hotels and Leisure	Yes
17	Healthcare	Yes
18	Bridging and/or Mezzanine Finance	No

Baltic Property Finance Plc

28 Savile Row
London
W1S 2EU

Tel 020 7025 1700

Contact
Robert Black

Criteria

1	Maximum Loan to Value	100%
2	Benchmark Lending Rate	Base
3	Margin Spread	Varies
4	Fixed	No
5	Minimum Loan	£500,000
6	Maximum Loan	£12,000,000
7	Arrangement Fee Spread	Varies
8	Minimum Length of Lease Required	Varies
9	Usual Length of Loan	1–3 years
10	Investment	No
11	Refinance	Yes
12	Owner Occupier	Yes
13	Residential Investment	No
14	Development Finance	Yes
15	Residential Development	No
16	Hotels and Leisure	No
17	Healthcare	No
18	Bridging and/or Mezzanine Finance	Yes

Julius Baer International Ltd

64 St James's Street
London
SW1A 1NF

Tel 020 7166 8100
Fax 020 7629 5931
www.juliusbaer.com

Contact
Charlotte Vilen
Private Banking Services &
Relationship Management

Criteria

1	Maximum Loan to Value	85%
2	Benchmark Lending Rate	3 month Libor
3	Margin Spread	1.5–2.0%
4	Fixed	No
5	Minimum Loan	£1,000,000
6	Maximum Loan	£5,000,000
7	Arrangement Fee Spread	0.5–1.0%
8	Minimum Length of Lease Required	30 years
9	Usual Length of Loan	3–5 years
10	Investment	Yes
11	Refinance	Yes
12	Owner Occupier	Yes
13	Residential Investment	Yes
14	Development Finance	No
15	Residential Development	Yes
16	Hotels and Leisure	Yes
17	Healthcare	No
18	Bridging and/or Mezzanine Finance	Yes

Bank Leumi (UK) Plc

20 Stratford Place
London
W1C 1BG

Tel 020 7907 8000
Fax 020 7908 8035
www.bankleumi.co.uk

Contact
David Griffiths
020 7907 8116
dgriffiths@bankleumi.co.uk

Malcolm Bloom
020 7907 8184
mbloom@bankleumi.co.uk

Criteria

1	Maximum Loan to Value	85%
2	Benchmark Lending Rate	Libor
3	Margin Spread	2–3%
4	Fixed	Possibly
5	Minimum Loan	£1,000,000
6	Maximum Loan	£15,000,000
7	Arrangement Fee Spread	1%
8	Minimum Length of Lease Required	N/A
9	Usual Length of Loan	Up to 3 years
10	Investment	Sometimes
11	Refinance	Yes
12	Owner Occupier	Possibly
13	Residential Investment	Yes
14	Development Finance	Yes
15	Residential Development	Yes
16	Hotels and Leisure	Yes
17	Healthcare	No
18	Bridging and/or Mezzanine Finance	Yes

Remarks

Short-term bridging and enablement finance, primarily for developments or breakup. Commercial developments should normally be pre-let.

Bank of East Asia

75 Shaftesbury Avenue
London
W1D 5BB

Tel 020 7734 3434
Fax 020 7734 0523
www.hkbea.co.uk

Contact
Kenneth Tam
info@hkbea.co.uk

Mr Chow York-wai, Joseph
General Manager

Criteria

1	Maximum Loan to Value	85%
2	Benchmark Lending Rate	Bank base/3 month Libor
3	Margin Spread	1.0–3.0%
4	Fixed	Yes
5	Minimum Loan	£250,000
6	Maximum Loan	£5,000,000
7	Arrangement Fee Spread	0.5%
8	Minimum Length of Lease Required	Varies
9	Usual Length of Loan	10–15 years
10	Investment	Yes
11	Refinance	Yes
12	Owner Occupier	Yes
13	Residential Investment	Yes
14	Development Finance	No
15	Residential Development	Yes
16	Hotels and Leisure	Yes
17	Healthcare	Yes
18	Bridging and/or Mezzanine Finance	No

Bank of Ireland

Eastcheap Court, 2nd Floor
11 Philpot Lane
London
EC3M 8BA

Tel 020 7560 8400
Fax 020 7626 2405
www.bank-of-ireland.co.uk

Contact
Jason Presence
Head of Property Finance UK
020 7560 8448
jason.presence@boiuk.com

Criteria

1	Maximum Loan to Value	90%
2	Benchmark Lending Rate	Libor
3	Margin Spread	0.5–1.0%
4	Fixed	Yes
5	Minimum Loan	£500,000
6	Maximum Loan	None
7	Arrangement Fee Spread	0.5–1.0%
8	Minimum Length of Lease Required	35 years
9	Usual Length of Loan	3–7 years
10	Investment	Yes
11	Refinance	Yes
12	Owner Occupier	No
13	Residential Investment	Yes
14	Development Finance	Yes
15	Residential Development	Yes
16	Hotels and Leisure	Yes
17	Healthcare	Yes
18	Bridging and/or Mezzanine Finance	Yes

Bank of Scotland (Centrebank)

Corporate Banking
Teviot House
41 South Gyle Crescent
Edinburgh
EH12 9BG

Tel 0131 442 5080
Fax 0131 317 5707
www.bankofscotland.co.uk

Contact
Gordon Smith
Alaistair Campbell

Criteria

1	Maximum Loan to Value	85%
2	Benchmark Lending Rate	Base rate
3	Margin Spread	1.0–2.0%
4	Fixed	Yes
5	Minimum Loan	£250,000
6	Maximum Loan	None
7	Arrangement Fee Spread	0.5–1.0%
8	Minimum Length of Lease Required	25 years plus length of loan
9	Usual Length of Loan	5–25 years or retirement
10	Investment	Yes
11	Refinance	Yes
12	Owner Occupier	Yes
13	Residential Investment	No
14	Development Finance	No
15	Residential Development	No
16	Hotels and Leisure	Yes
17	Healthcare	Yes
18	Bridging and/or Mezzanine Finance	No

Bank of Tokyo — Mitsubishi

Finsbury Circus House
12-15 Finsbury Circus
London
EC2M 7BT

Tel 020 7577 1717
020 7588 1111
Fax 020 7577 1759
www.btm.co.uk

Contact
Alistair Mackie

Criteria

1	Maximum Loan to Value	80%
2	Benchmark Lending Rate	3 month Libor
3	Margin Spread	0.75–2.0%
4	Fixed	Yes
5	Minimum Loan	£5,000,000
6	Maximum Loan	None
7	Arrangement Fee Spread	0.25–0.75%
8	Minimum Length of Lease Required	Flexible
9	Usual Length of Loan	5–10 years
10	Investment	Yes
11	Refinance	Yes
12	Owner Occupier	Yes
13	Residential Investment	No
14	Development Finance	Yes
15	Residential Development	No
16	Hotels and Leisure	No
17	Healthcare	No
18	Bridging and/or Mezzanine Finance	Yes

Bankers Trust

1 Appold Street
Broadgate
London
EC2A 2HE

Tel 020 7982 2500/1593
Fax 020 7982 1593/3308

Contact
Alistair Dixon
020 7982 3714

Richard Mully
020 7982 2745

Criteria

1	Maximum Loan to Value	Providers of mezzanine debt
2	Benchmark Lending Rate	Base
3	Margin Spread	Varied
4	Fixed	No
5	Minimum Loan	£3,000,000
6	Maximum Loan	£25,000,000
7	Arrangement Fee Spread	Varies
8	Minimum Length of Lease Required	None
9	Usual Length of Loan	3–5 years
10	Investment	Yes
11	Refinance	Yes
12	Owner Occupier	Yes
13	Residential Investment	No
14	Development Finance	No
15	Residential Development	No
16	Hotels and Leisure	No
17	Healthcare	No
18	Bridging and/or Mezzanine Finance	Yes

Remarks

Bankers Trust aims to adopt a structured approach to the provision of mezzanine finance to enable the borrower to achieve enhanced returns. Providers of mezzanine and equity debt only.

BNP Paribas

10 Harewood Avenue
London
NW1 6AA

Tel 020 7595 5254
Fax 020 7595 6192
www.bnpparibas.co.uk

Contact

Julian van Kan
Head of Loan Syndications
and Trading
020 7595 6591

Criteria

1	Maximum Loan to Value	N/A
2	Benchmark Lending Rate	N/A
3	Margin Spread	N/A
4	Fixed	N/A
5	Minimum Loan	N/A
6	Maximum Loan	N/A
7	Arrangement Fee Spread	N/A
8	Minimum Length of Lease Required	N/A
9	Usual Length of Loan	N/A
10	Investment	N/A
11	Refinance	N/A
12	Owner Occupier	N/A
13	Residential Investment	N/A
14	Development Finance	N/A
15	Residential Development	N/A
16	Hotels and Leisure	N/A
17	Healthcare	N/A
18	Bridging and/or Mezzanine Finance	N/A

Remarks

The bank still administers its existing loan book but is not writing new business.

Barclays Bank Plc

54 Lombard Street
London
EC3P 3AH

Tel 020 7699 2755
Fax 020 7699 3882
www.barclays.co.uk

Contact
Andy Chaplin
andy.chaplin@barclays.co.uk

Criteria

1	Maximum Loan to Value	90% for mortgages. Special scheme available for doctors. 80% for property investment.
2	Benchmark Lending Rate	Base/Libor
3	Margin Spread	1.0–5.0% normal
4	Fixed	Yes
5	Minimum Loan	£25,000
6	Maximum Loan	Unlimited
7	Arrangement Fee Spread	Negotiable
8	Minimum Length of Lease Required	Negotiable
9	Usual Length of Loan	Up to 25 years commercial mortgage Up to 20 years residential property investment Up to 10 years commercial property investment
10	Investment	Yes
11	Refinance	Yes
12	Owner Occupier	Yes
13	Residential Investment	Yes
14	Development Finance	Yes
15	Residential Development	Yes
16	Hotels and Leisure	Yes
17	Healthcare	Yes
18	Bridging and/or Mezzanine Finance	Yes

Barclays Property Team

54 Lombard Street
London
EC3P 3AH

Tel 020 7699 5000
Fax 020 7699 3882
www.barclays.co.uk/commercial

Contact
John Dewberry
Head of Property Team
020 7116 4491
john.dewberry@barclayscorporate.c

David Cooper
Head of Structured Project Finance
david.cooper1@barclays.com

Criteria

1	Maximum Loan to Value	90%
2	Benchmark Lending Rate	Libor
3	Margin Spread	0.6–2.0%
4	Fixed	Yes
5	Minimum Loan	£2,000,000
6	Maximum Loan	None
7	Arrangement Fee Spread	0.4–1.0%
8	Minimum Length of Lease Required	Varies
9	Usual Length of Loan	Up to 3 years for development finance Up to 10 years for investment finance
10	Investment	Yes
11	Refinance	Yes
12	Owner Occupier	Yes
13	Residential Investment	Yes
14	Development Finance	Yes
15	Residential Development	Yes
16	Hotels and Leisure	Yes
17	Healthcare	No
18	Bridging and/or Mezzanine Finance	Yes

Barnsley Building Society

Permanent Building
Regent Street
Barnsley
S70 2EH

Tel 01226 733999
Fax 01226 214925
www.barnsley-bs.co.uk
maps.process@barnsley-bs.co.uk

Contact
Hazel Womersley
Lending Manager

Diane Ryder
Supervisor

Clare Hawkins
Supervisor

Criteria

1	Maximum Loan to Value	95%
2	Benchmark Lending Rate	Standard variable rate
3	Margin Spread	Up to 2%
4	Fixed	Yes
5	Minimum Loan	£50,000
6	Maximum Loan	£1,000,000
7	Arrangement Fee Spread	1.0%
8	Minimum Length of Lease Required	Mortgage term plus 50 years
9	Usual Length of Loan	5–25 years
10	Investment	Yes
11	Refinance	Yes
12	Owner Occupier	Yes
13	Residential Investment	No
14	Development Finance	No
15	Residential Development	No
16	Hotels and Leisure	No
17	Healthcare	No
18	Bridging and/or Mezzanine Finance	No

Bath Building Society

20 Charles Street
Bath
BA1 1HY

Tel 01225 423271
Fax 01225 424590
www.bibs.co.uk
mortgages@bibs.co.uk

Contact
Steve Matthens
Sales Manager

Chris Powell
Senior Underwriter

Criteria

1	Maximum Loan to Value	95% — residential 75% — semi-commercial 67% — wholly commercial
2	Benchmark Lending Rate	Standard variable rate
3	Margin Spread	1.5–2.0%
4	Fixed	No
5	Minimum Loan	£40,000
6	Maximum Loan	By agreement
7	Arrangement Fee Spread	1.0%
8	Minimum Length of Lease Required	50 years
9	Usual Length of Loan	5–40 years
10	Investment	Yes
11	Refinance	Yes
12	Owner Occupier	Yes
13	Residential Investment	Yes
14	Development Finance	Yes
15	Residential Development	Sometimes
16	Hotels and Leisure	Yes
17	Healthcare	Yes
18	Bridging and/or Mezzanine Finance	No

Remarks

50-mile radius of Bath.

Bear Stearns

One Canada Square
London
E14 5AD

Tel 020 7516 6000
www.bearstearns.com

Contact
Charles Pitt
020 7516 6000

Criteria

1	Maximum Loan to Value	100%
2	Benchmark Lending Rate	Libor
3	Margin Spread	Varies
4	Fixed	Yes
5	Minimum Loan	£10,000,000
6	Maximum Loan	None
7	Arrangement Fee Spread	0.5–1.0%
8	Minimum Length of Lease Required	Varies
9	Usual Length of Loan	Varies
10	Investment	Yes
11	Refinance	Yes
12	Owner Occupier	No
13	Residential Investment	No
14	Development Finance	Yes
15	Residential Development	No
16	Hotels and Leisure	Yes
17	Healthcare	Yes
18	Bridging and/or Mezzanine Finance	Yes

Remarks

Bear Stearns is going through profound changes so the criteria may change, but was correct as of 16 March 2008.

Berlin Hyp

1 Crown Court
66 Cheapside
London
EC2V 6LR

Fax 0207 572 6219
www.berlinhyp.de

Contact
Elizabeth Edwards
Head of Marketing
020 7572 6493

Criteria

1	Maximum Loan to Value	90%
2	Benchmark Lending Rate	Libor
3	Margin Spread	1–2.5%
4	Fixed	Yes
5	Minimum Loan	£2,000,000
6	Maximum Loan	£10,000,000
7	Arrangement Fee Spread	0.5–1.0%
8	Minimum Length of Lease Required	60 years
9	Usual Length of Loan	5–10 years
10	Investment	Yes
11	Refinance	Yes
12	Owner Occupier	No
13	Residential Investment	Yes
14	Development Finance	Yes
15	Residential Development	No
16	Hotels and Leisure	No
17	Healthcare	No
18	Bridging and/or Mezzanine Finance	No

BHF-Bank AG

61 Queen Street
London
EC4R 1AE

Tel 020 7634 2300
Fax 020 7248 5739
www.bhf-bank.com

Contact
Salman Hasan
Tony Whetstone

Criteria

1	Maximum Loan to Value	80%
2	Benchmark Lending Rate	Libor
3	Margin Spread	Varies
4	Fixed	Yes
5	Minimum Loan	£1,000,000
6	Maximum Loan	£10,000,000
7	Arrangement Fee Spread	Negotiable
8	Minimum Length of Lease Required	Varies
9	Usual Length of Loan	1–5 years
10	Investment	Yes
11	Refinance	Yes
12	Owner Occupier	No
13	Residential Investment	No
14	Development Finance	Yes
15	Residential Development	Yes
16	Hotels and Leisure	Yes
17	Healthcare	Yes
18	Bridging and/or Mezzanine Finance	Yes

Birmingham Midshires

Business & Professional Finance
PO Box 81
Pendeford Business Park
Wobaston Road
Wolverhampton
WV9 5HZ

Tel 0845 300 2627
www.askbm.co.uk

Contact
Nigel Stockton
lending@birminghammidshires.co.uk

Criteria

1	Maximum Loan to Value	85%
2	Benchmark Lending Rate	3 month Libor
3	Margin Spread	1–2.5%
4	Fixed	Yes
5	Minimum Loan	£100,000
6	Maximum Loan	£10,000,000
7	Arrangement Fee Spread	Negotiable
8	Minimum Length of Lease Required	N/A
9	Usual Length of Loan	25 years max 5 years min
10	Investment	Yes
11	Refinance	Yes
12	Owner Occupier	Yes
13	Residential Investment	No
14	Development Finance	Yes
15	Residential Development	No
16	Hotels and Leisure	Yes
17	Healthcare	Yes
18	Bridging and/or Mezzanine Finance	No

Remarks

Now part of Halifax Group.

Bradford & Bingley Plc

Commercial Lending
PO Box 88
Crossflats
Bingley
West Yorkshire
BD16 2UA

Tel 0127 4554188
Fax 01274 544195
www.bbg.co.uk

Contact
Liz Walker
liz.walker@bbg.co.uk

Emma Bates

Criteria

1	Maximum Loan to Value	85%
2	Benchmark Lending Rate	Libor
3	Margin Spread	Negotiable
4	Fixed	Up to 25 years
5	Minimum Loan	£250,000
6	Maximum Loan	None
7	Arrangement Fee Spread	0.25–1.0%
8	Minimum Length of Lease Required	N/A
9	Usual Length of Loan	2–25 years
10	Investment	Yes
11	Refinance	Yes
12	Owner Occupier	No
13	Residential Investment	No
14	Development Finance	Yes
15	Residential Development	Yes
16	Hotels and Leisure	No
17	Healthcare	No
18	Bridging and/or Mezzanine Finance	No

Bristol & West Property Finance

PO Box 28
One Temple Quay
Bristol
BS99 7AX

Tel 0117 943 2531
Fax 0117 929 3814
www.bristol-west.co.uk

Contact
James Penman
Head of Bristol & West
Property Finance
0117 943 2455
james.penman@boiuk.com

Criteria

1	Maximum Loan to Value	Negotiable
2	Benchmark Lending Rate	3 month Libor
3	Margin Spread	Negotiable
4	Fixed	Yes
5	Minimum Loan	£1,000,000
6	Maximum Loan	£50,000,000
7	Arrangement Fee Spread	Negotiable
8	Minimum Length of Lease Required	Negotiable
9	Usual Length of Loan	3–15 years
10	Investment	Yes
11	Refinance	Yes
12	Owner Occupier	Yes
13	Residential Investment	Yes
14	Development Finance	Yes
15	Residential Development	Yes
16	Hotels and Leisure	Yes
17	Healthcare	Yes
18	Bridging and/or Mezzanine Finance	Yes

Britannia Commercial Lending

Britannia House 2nd floor
10 Town Road
Hanley
Stoke-on-Trent
ST1 2QQ

Tel 01782 408500
Fax 01782 408559
www.britannia.co.uk

Contact

Steve Goldstraw
Managing Director
steve.goldstraw@britannia.co.uk

Mike Rogers
Strategy Manager
mike.rogers@britannia.co.uk

Dave Dono
Strategy Manager
david.dono@britannia.co.uk

Martin Latham
Strategy Manager
martin.latham@britannia.co.uk

Criteria

1	Maximum Loan to Value	85%
2	Benchmark Lending Rate	Bespoke
3	Margin Spread	Bespoke
4	Fixed	Bespoke
5	Minimum Loan	£1,000,000
6	Maximum Loan	None
7	Arrangement Fee Spread	Bespoke
8	Minimum Length of Lease Required	1–30 years
9	Usual Length of Loan	1–30 years
10	Investment	Yes
11	Refinance	Yes
12	Owner Occupier	Yes
13	Residential Investment	Yes
14	Development Finance	No
15	Residential Development	No
16	Hotels and Leisure	Yes
17	Healthcare	Yse
18	Bridging and/or Mezzanine Finance	Bridging finance — No Mezzanine finance — Yes

British Arab Commercial Bank Ltd

8-10 Mansion House Place
London
EC4N 8BJ

Tel 020 7648 7777
Fax 020 7648 7774
www.bacb.co.uk

Contact
Geoff Duncanson
Head of Lending
020 7972 6902
enquiries@bacb.co.uk

Sue Tanner
Syndications
020 7972 6933

Criteria

1	Maximum Loan to Value	75%
2	Benchmark Lending Rate	Libor
3	Margin Spread	1%+
4	Fixed	Yes
5	Minimum Loan	£1,000,000
6	Maximum Loan	£15,000,000
7	Arrangement Fee Spread	Negotiable
8	Minimum Length of Lease Required	N/A
9	Usual Length of Loan	5 years
10	Investment	Yes
11	Refinance	Yes
12	Owner Occupier	No
13	Residential Investment	No
14	Development Finance	No
15	Residential Development	No
16	Hotels and Leisure	No
17	Healthcare	No
18	Bridging and/or Mezzanine Finance	No

Buckinghamshire Building Society

High Street
Chalfont St. Giles
Bucks
HP8 4QB

Tel 01494 879500
Fax 01494 876256
www.bucksbuildingsociety.co.uk

Contact
Mortgage Underwriting Team
01753 482100
info@bucksbuildingsociety.co.uk

Criteria

1	Maximum Loan to Value	80%
2	Benchmark Lending Rate	Base
3	Margin Spread	1.0–2.0%
4	Fixed	Yes
5	Minimum Loan	£50,000
6	Maximum Loan	£500,000
7	Arrangement Fee Spread	1.0–3.0%
8	Minimum Length of Lease Required	75 years
9	Usual Length of Loan	5–25 years
10	Investment	Yes
11	Refinance	Yes
12	Owner Occupier	Yes
13	Residential Investment	Yes
14	Development Finance	No
15	Residential Development	No
16	Hotels and Leisure	No
17	Healthcare	Yes
18	Bridging and/or Mezzanine Finance	No

Butterfield Private Bank

99 Gresham Street
London
EC2V 7NG

Tel 020 7776 6700
Fax 020 7776 6701
www.butterfieldbank.com

Contact
Mark Lean
Property Finance

Criteria

1	Maximum Loan to Value	85%
2	Benchmark Lending Rate	Base
3	Margin Spread	1.0–2.0%
4	Fixed	Yes
5	Minimum Loan	£250,000
6	Maximum Loan	£5,000,000
7	Arrangement Fee Spread	0.5–1.0%
8	Minimum Length of Lease Required	40 years
9	Usual Length of Loan	1–5 years
10	Investment	Yes
11	Refinance	Yes
12	Owner Occupier	Yes
13	Residential Investment	Yes
14	Development Finance	Yes
15	Residential Development	Yes
16	Hotels and Leisure	No
17	Healthcare	No
18	Bridging and/or Mezzanine Finance	No

Remarks

Formerly Leopold Joseph & Sons Ltd.

Cambridge Building Society

PO Box 232
51 Newmarket Road
Cambridge
CB5 8FF

Tel 01223 727727
Fax 01223 727700
www.cambridgebs.co.uk

Contact
Tony Millington
Head of Finance

Criteria

1	Maximum Loan to Value	75%
2	Benchmark Lending Rate	Own residential base rate
3	Margin Spread	Base–2.5%
4	Fixed	No
5	Minimum Loan	£25,000
6	Maximum Loan	£1,000,000
7	Arrangement Fee Spread	0.5–1.0%
8	Minimum Length of Lease Required	25 years plus length of loan
9	Usual Length of Loan	5–25 years
10	Investment	Yes
11	Refinance	Yes
12	Owner Occupier	Yes
13	Residential Investment	Yes
14	Development Finance	Yes
15	Residential Development	Yes
16	Hotels and Leisure	Yes
17	Healthcare	Yes
18	Bridging and/or Mezzanine Finance	No

Remarks

Commercial lending only provided on property within a 30-mile radius of Cambridge.

Canada Life Ltd

Canada Life Place
Potters Bar
Hertfordshire
EN6 5BA

Tel 08456 06 07 08
Fax 01707 42 31 30
www.canadalife.co.uk

Contact
Eleni Poullides

Criteria

1	Maximum Loan to Value	90%
2	Benchmark Lending Rate	Swap
3	Margin Spread	80bp
4	Fixed	Yes, 10–30 years
5	Minimum Loan	£5,000,000
6	Maximum Loan	None
7	Arrangement Fee Spread	0.25–0.5%
8	Minimum Length of Lease Required	10 years
9	Usual Length of Loan	10–30 years
10	Investment	Yes
11	Refinance	Yes
12	Owner Occupier	No
13	Residential Investment	No
14	Development Finance	No
15	Residential Development	Yes
16	Hotels and Leisure	Yes
17	Healthcare	No
18	Bridging and/or Mezzanine Finance	No

Remarks

The company grants long-term fixed interest loans on good quality commercial property. Five-year floating rate loans are also available.

Capital Home Loans Ltd

Admiral House
Harlington Way
Fleet
Hampshire
GU51 4YA

Tel 01252 812271
Fax 01252 811826
www.chlmortgages.co.uk

Contact
David Guinane
Managing Director

Criteria

1	Maximum Loan to Value	100%
2	Benchmark Lending Rate	Libor/Bank base
3	Margin Spread	Varies
4	Fixed	Yes
5	Minimum Loan	£25,000
6	Maximum Loan	£1,000,000
7	Arrangement Fee Spread	0.25–1.0%
8	Minimum Length of Lease Required	Varies
9	Usual Length of Loan	25 years
10	Investment	No
11	Refinance	Residential only
12	Owner Occupier	Yes
13	Residential Investment	Yes
14	Development Finance	No
15	Residential Development	No
16	Hotels and Leisure	No
17	Healthcare	No
18	Bridging and/or Mezzanine Finance	No

Remarks

The residential investment side focuses on housing associations.

Charterhouse Property Finance

1 Paternoster Row
St Pauls
London
EC4M 7DH

Tel 020 7248 4000
Fax 020 7246 2110

Contact
Andrew Moffat

Criteria

1	Maximum Loan to Value	100%
2	Benchmark Lending Rate	3 month Libor
3	Margin Spread	0.75–1.5%
4	Fixed	Yes
5	Minimum Loan	£2,500,000
6	Maximum Loan	None
7	Arrangement Fee Spread	0.25–1.5%
8	Minimum Length of Lease Required	Varies
9	Usual Length of Loan	3 months–5 years
10	Investment	Yes
11	Refinance	Yes
12	Owner Occupier	Yes
13	Residential Investment	Yes
14	Development Finance	Yes
15	Residential Development	Yes
16	Hotels and Leisure	Yes
17	Healthcare	No
18	Bridging and/or Mezzanine Finance	Yes

Remarks

Prepared to actively participate in some transactions. Specialises in high loan to value and also offers mezzanine facilities.

Chase Manhattan, JPMorgan

60 Victoria Embankment
London
EC4Y 0JP

Tel 020 7600 2300

Contact
Christopher King
Director

Criteria

1	Maximum Loan to Value	100%
2	Benchmark Lending Rate	Libor
3	Margin Spread	Varies
4	Fixed	Yes
5	Minimum Loan	£20,000,000
6	Maximum Loan	None
7	Arrangement Fee Spread	Negotiable
8	Minimum Length of Lease Required	Varies
9	Usual Length of Loan	Varies
10	Investment	Yes
11	Refinance	Yes
12	Owner Occupier	No
13	Residential Investment	N/A
14	Development Finance	N/A
15	Residential Development	Yes
16	Hotels and Leisure	Yes
17	Healthcare	Yes
18	Bridging and/or Mezzanine Finance	Yes

Remarks

Primarily funding for larger leisure deals.

Chelsea Building Society

Thirlestaine Hall
Thirlestaine Road
Cheltenham
Gloucester
GL53 7AL

Tel 01242 271271
Fax 01242 571441
www.thechelsea.co.uk

Contact
David Richards
Mortgage Practice Manager
01242 271428

Criteria

1	Maximum Loan to Value	95%
2	Benchmark Lending Rate	Standard variable rate 6.34%
3	Margin Spread	N/A
4	Fixed	N/A
5	Minimum Loan	£20,000
6	Maximum Loan	£1,000,000
7	Arrangement Fee Spread	N/A
8	Minimum Length of Lease Required	55 years
9	Usual Length of Loan	5–25 years
10	Investment	N/A
11	Refinance	N/A
12	Owner Occupier	Yse
13	Residential Investment	Yes
14	Development Finance	No
15	Residential Development	No
16	Hotels and Leisure	N/A
17	Healthcare	N/A
18	Bridging and/or Mezzanine Finance	No

Cheltenham & Gloucester Plc

Barnet Way
Barnwood
Gloucester
GL4 3RL

Tel 01452 372372
Fax 01452 375879
www.cheltglos.co.uk

Contact
Brian Harwood

Criteria

1	Maximum Loan to Value	N/A
2	Benchmark Lending Rate	N/A
3	Margin Spread	N/A
4	Fixed	N/A
5	Minimum Loan	N/A
6	Maximum Loan	N/A
7	Arrangement Fee Spread	N/A
8	Minimum Length of Lease Required	N/A
9	Usual Length of Loan	N/A
10	Investment	N/A
11	Refinance	N/A
12	Owner Occupier	N/A
13	Residential Investment	N/A
14	Development Finance	N/A
15	Residential Development	N/A
16	Hotels and Leisure	N/A
17	Healthcare	N/A
18	Bridging and/or Mezzanine Finance	N/A

Remarks

The commercial department is only lending to existing clients.

Chesham Building Society

12 Market Square
Chesham
Bucks
HP5 1ER

Tel 0800 068 4784
Fax 01494 778399
www.thechesham.co.uk

Contact
Richard Krasucki
Finance Manager
mortgages@thechesham.co.uk

Criteria

1	Maximum Loan to Value	85%
2	Benchmark Lending Rate	Base
3	Margin Spread	1.0–2.0%
4	Fixed	Yes
5	Minimum Loan	£125,000
6	Maximum Loan	£1,000,000
7	Arrangement Fee Spread	1–1.5%
8	Minimum Length of Lease Required	50 years
9	Usual Length of Loan	5–25 years
10	Investment	Yes
11	Refinance	Yes
12	Owner Occupier	Yes
13	Residential Investment	Yes
14	Development Finance	No
15	Residential Development	No
16	Hotels and Leisure	Yes
17	Healthcare	Yes
18	Bridging and/or Mezzanine Finance	No

Cheshire Building Society

Castle Street
Macclesfield
Cheshire
SK11 6AF

Tel 01625 613612
Fax 06125 617147
www.thecheshire.co.uk

Contact
Andy Russell
Finance Director

Criteria

1	Maximum Loan to Value	75%
2	Benchmark Lending Rate	Standard variable rate
3	Margin Spread	0.5%
4	Fixed	Yes
5	Minimum Loan	£15,001
6	Maximum Loan	£1,000,000
7	Arrangement Fee Spread	1.0%
8	Minimum Length of Lease Required	None
9	Usual Length of Loan	5–20 years
10	Investment	Yes
11	Refinance	Yes
12	Owner Occupier	Yes
13	Residential Investment	Yes
14	Development Finance	No
15	Residential Development	No
16	Hotels and Leisure	Yes
17	Healthcare	Yes
18	Bridging and/or Mezzanine Finance	No

Citibank International Plc

Citibank Business & Commercial
St Martins House
1 Hammersmith Grove
London
W6 0NY

Tel 020 7500 3127
Fax 020 7500 3591

Contact
Alaistair Caie

Criteria

1	Maximum Loan to Value	90% (100% healthcare)
2	Benchmark Lending Rate	Bank base
3	Margin Spread	1.0–3.5%
4	Fixed	Sometimes
5	Minimum Loan	£250,000
6	Maximum Loan	None
7	Arrangement Fee Spread	1%
8	Minimum Length of Lease Required	99 years
9	Usual Length of Loan	5–25 years
10	Investment	Yes
11	Refinance	Yes
12	Owner Occupier	Yes, healthcare only
13	Residential Investment	Yes
14	Development Finance	No
15	Residential Development	No
16	Hotels and Leisure	Yes
17	Healthcare	Yes
18	Bridging and/or Mezzanine Finance	No

City Group

Victoria Plaza
111 Buckingham Palace Road
London
SW1W 0SB

Tel 020 7721 2000
Fax 020 7721 2834

Contact
Ken Beisser
Robert Balick
Nigel King

Criteria

1	Maximum Loan to Value	100% subject to status or income stream
2	Benchmark Lending Rate	Base
3	Margin Spread	Varies
4	Fixed	No
5	Minimum Loan	£10,000,000
6	Maximum Loan	None
7	Arrangement Fee Spread	Varies
8	Minimum Length of Lease Required	None
9	Usual Length of Loan	Generally short-term lending
10	Investment	Yes
11	Refinance	Yes
12	Owner Occupier	Yes
13	Residential Investment	No
14	Development Finance	Yes
15	Residential Development	Yes
16	Hotels and Leisure	Yes
17	Healthcare	No
18	Bridging and/or Mezzanine Finance	Yes

Close Property Finance Ltd

10 Crown Place
London
EC2A 4FT

Tel 020 7655 3100
Fax 020 7247 1205
www.closepropertyfinance.com

Contact

Robert Orr
Director of Property Finance
020 7655 3412
rorr@closebrothers.co.uk

Rowland Thomas
Director
020 7655 3775

Simon Powell
Assistant Director
020 7655 3438

Criteria

1	Maximum Loan to Value	75%
2	Benchmark Lending Rate	Libor
3	Margin Spread	2.0–4.0%
4	Fixed	Yes
5	Minimum Loan	£200,000
6	Maximum Loan	£20,000,000
7	Arrangement Fee Spread	1.0%
8	Minimum Length of Lease Required	No minimum — provided spread
9	Usual Length of Loan	1–5 years
10	Investment	Yes
11	Refinance	Yes
12	Owner Occupier	No
13	Residential Investment	Yes
14	Development Finance	Yes
15	Residential Development	Yes
16	Hotels and Leisure	No
17	Healthcare	No
18	Bridging and/or Mezzanine Finance	Yes — bridging No — mezannine

Remarks

For loans of £200,000–£3,000,000 contact: Daniel Joyce Tel: 020 7655 3770, Lee Daniel Tel: 020 7655 3783, Richard Hemmings Tel: 020 7655 3780. For loans of £3,000,000–£20,000,000 see under Contact. Lends on development with repayment date. Also residential and commercial development. The latter predominantly pre-let. Lends on secondary investment portfolios with good interest cover.

Clydesdale Bank

30 St Vincent Place
Glasgow
G1 2HL

Tel 0141 248 7070
www.cbonline.co.uk

Contact

Contact the commercial mortgage manager at the nearest branch concerned.

Criteria

1	Maximum Loan to Value	80%
2	Benchmark Lending Rate	Bank/Libor
3	Margin Spread	1.0–5.0%
4	Fixed	Sometimes
5	Minimum Loan	£25,000
6	Maximum Loan	None
7	Arrangement Fee Spread	0.5–1.0%
8	Minimum Length of Lease Required	20 years
9	Usual Length of Loan	1–20 years
10	Investment	Yes
11	Refinance	Yes
12	Owner Occupier	Yes
13	Residential Investment	Yes
14	Development Finance	Yes
15	Residential Development	Yes
16	Hotels and Leisure	Yes
17	Healthcare	Yes
18	Bridging and/or Mezzanine Finance	No

Commercial Acceptances

Hall Studio
23B Grove End Road
London
NW8 9BP

Tel 020 7289 3326
Fax 020 7286 9280

Contact
Neville Freed
L B Brown
D A Hertz

Criteria

1	Maximum Loan to Value	70%
2	Benchmark Lending Rate	Bank base
3	Margin Spread	3.0%
4	Fixed	No
5	Minimum Loan	£25,000
6	Maximum Loan	£4,000,000
7	Arrangement Fee Spread	1%, refundable
8	Minimum Length of Lease Required	60 years
9	Usual Length of Loan	1 day–6 months
10	Investment	Yes
11	Refinance	Yes
12	Owner Occupier	Yes
13	Residential Investment	Yes
14	Development Finance	Pre-let only
15	Residential Development	Yes
16	Hotels and Leisure	No
17	Healthcare	No
18	Bridging and/or Mezzanine Finance	Yes

Remarks

Specialises in bridging finance from 1 day to 6 months. Likes deals where a quick reaction is required.

Co-operative Bank

PO Box 101
1 Balloon Street
Manchester
M60 4EP

Tel 0161 8295349
www.co-operativebank.co.uk

Contact
Commercial Mortgage Manager
at nearest branch
0845 721 2212

Criteria

1	Maximum Loan to Value	75%
2	Benchmark Lending Rate	Bank base
3	Margin Spread	2.5–5.0%
4	Fixed	Yes
5	Minimum Loan	Varies
6	Maximum Loan	None
7	Arrangement Fee Spread	1.0–1.5%
8	Minimum Length of Lease Required	35 years
9	Usual Length of Loan	1–10 years
10	Investment	Yes
11	Refinance	Yes
12	Owner Occupier	Yes
13	Residential Investment	No
14	Development Finance	No
15	Residential Development	No
16	Hotels and Leisure	Yes
17	Healthcare	Yes
18	Bridging and/or Mezzanine Finance	No

Coutts & Co

440 Strand
London
WC2R 0QS

Tel 020 7957 2424
Fax 020 7753 1050
www.coutts.com

Contact
Commercial Mortgage Manager
at nearest branch

Criteria

1	Maximum Loan to Value	90%
2	Benchmark Lending Rate	Libor
3	Margin Spread	0.5–1.0%
4	Fixed	Yes
5	Minimum Loan	£750,000
6	Maximum Loan	None
7	Arrangement Fee Spread	0.5–1.0%
8	Minimum Length of Lease Required	70 years
9	Usual Length of Loan	Up to 25 years
10	Investment	Yes
11	Refinance	Yes
12	Owner Occupier	Yes
13	Residential Investment	Yes
14	Development Finance	Yes
15	Residential Development	Yes
16	Hotels and Leisure	No
17	Healthcare	Yes
18	Bridging and/or Mezzanine Finance	Yes

Coventry Building Society

Economic House
PO Box 9
High Street
Coventry
CV1 5QN

Tel 0845 7665522
www.coventrybuildingsociety.co.uk

Contact
Rob Green
Head of Finance

Criteria

1	Maximum Loan to Value	85%
2	Benchmark Lending Rate	Building society base rate
3	Margin Spread	Varies
4	Fixed	Yes
5	Minimum Loan	£25,000
6	Maximum Loan	£1,000,000
7	Arrangement Fee Spread	1.0%
8	Minimum Length of Lease Required	N/A
9	Usual Length of Loan	5–25 years
10	Investment	Yes
11	Refinance	No
12	Owner Occupier	No
13	Residential Investment	Yes
14	Development Finance	No
15	Residential Development	No
16	Hotels and Leisure	No
17	Healthcare	No
18	Bridging and/or Mezzanine Finance	No

Remarks

Lending in 50-mile radius of Coventry.

Crédit Agricole Indosuez

122 Leadenhall Street
London
EC2V 4QH

Tel 020 7971 4000
Fax 020 7971 4406

Contact
Nicola Crupton
Peter Stevens

Criteria

1	Maximum Loan to Value	75%
2	Benchmark Lending Rate	Libor
3	Margin Spread	1.0–2.0%
4	Fixed	Yes
5	Minimum Loan	£2,000,000
6	Maximum Loan	£15,000,000
7	Arrangement Fee Spread	0.25–0.75%
8	Minimum Length of Lease Required	10 years
9	Usual Length of Loan	1–5 years
10	Investment	Yes
11	Refinance	Yes
12	Owner Occupier	Sometimes
13	Residential Investment	No
14	Development Finance	No
15	Residential Development	No
16	Hotels and Leisure	No
17	Healthcare	No
18	Bridging and/or Mezzanine Finance	No

Remarks

Lending to existing clients only.

Credit Suisse

Asset Management
One Cabot Square
London
E14 4QJ

Tel 020 7883 7734
Fax 020 7888 1600
www.credit-suisse.co.uk

Contact
Real Estate Financing Team

Criteria

1	Maximum Loan to Value	85%
2	Benchmark Lending Rate	Libor
3	Margin Spread	0.5–1.0%
4	Fixed	Yes
5	Minimum Loan	£5,000,000
6	Maximum Loan	None
7	Arrangement Fee Spread	0.5–1.0%
8	Minimum Length of Lease Required	Varies
9	Usual Length of Loan	1–10 years
10	Investment	Yes
11	Refinance	Yes
12	Owner Occupier	No
13	Residential Investment	No
14	Development Finance	Yes
15	Residential Development	No
16	Hotels and Leisure	Yes
17	Healthcare	Yes
18	Bridging and/or Mezzanine Finance	Yes

Cumberland Building Society

Cumberland House
Castle Street
Carlisle
CA3 8RX

Tel 0845 6023462
www.cumberland.co.uk

Contact
Stephen Barringer
Commercial Lending Team

Bill Smith
Commercial Lending Team

Peter Temple
Operations Director
peter.temple@cumberland.co.uk

Criteria

1	Maximum Loan to Value	75%
2	Benchmark Lending Rate	Base
3	Margin Spread	1.0–2.0%
4	Fixed	Yes
5	Minimum Loan	£250,000
6	Maximum Loan	£850,000
7	Arrangement Fee Spread	1.0–3.00%
8	Minimum Length of Lease Required	70 years
9	Usual Length of Loan	5–15 years
10	Investment	Yes
11	Refinance	Yes
12	Owner Occupier	Yes
13	Residential Investment	Yes
14	Development Finance	No
15	Residential Development	No
16	Hotels and Leisure	No
17	Healthcare	No
18	Bridging and/or Mezzanine Finance	No

Darlington Building Society

Sentinel House
Lingfield Way
Darlington
Co Durham
DL1 4PR

Tel 01325 366366
Fax 01325 741000
www.darlington.co.uk
commercial@darlington.co.uk

Contact
Robert Murphy
Commercial Manager

Criteria

1	Maximum Loan to Value	75%
2	Benchmark Lending Rate	Building society standard variable rate
3	Margin Spread	0–1.5%
4	Fixed	Yes — minimum £500,000
5	Minimum Loan	£30,100
6	Maximum Loan	£1,000,000
7	Arrangement Fee Spread	0.25–1.0%
8	Minimum Length of Lease Required	To exceed mortgage term
9	Usual Length of Loan	5–25 years
10	Investment	Yes
11	Refinance	Yes
12	Owner Occupier	Yes
13	Residential Investment	Yes
14	Development Finance	No
15	Residential Development	No
16	Hotels and Leisure	Some
17	Healthcare	Some
18	Bridging and/or Mezzanine Finance	No

Remarks

South Durham/North Yorkshire area preferable but will consider England and Wales if coming from a known source.

Depfa Bank Plc

105 Wigmore Street
London
W1U 1QY

Tel 020 7290 8400
Fax 020 7495 0580
www.depfa.com

Contact
Francis Hazeel

James Campbell

Criteria

1	Maximum Loan to Value	100%
2	Benchmark Lending Rate	Cost of funds/Libor
3	Margin Spread	70–150bp
4	Fixed	Requirement for hedging
5	Minimum Loan	Investment (bipartite) £5,000,000 Investment (club) and speculative £10,000,000
6	Maximum Loan	N/A
7	Arrangement Fee Spread	30–125bp
8	Minimum Length of Lease Required	Varies
9	Usual Length of Loan	5–10 years
10	Investment	Yes
11	Refinance	Yes
12	Owner Occupier	Yes
13	Residential Investment	No
14	Development Finance	Yes
15	Residential Development	No
16	Hotels and Leisure	Yes
17	Healthcare	No
18	Bridging and/or Mezzanine Finance	—

Derbyshire Building Society

Duffield Hall
Duffield
Derby
DE56 1AG

Tel 01332 841000
Fax 01332 844666
www.thederbyshire.co.uk

Contact
George Elefthiriou
Head of Commercial Lending

Criteria

1	Maximum Loan to Value	80%
2	Benchmark Lending Rate	Building Society cost of funds
3	Margin Spread	1.0–3.0%
4	Fixed	Yes
5	Minimum Loan	£100,000
6	Maximum Loan	None
7	Arrangement Fee Spread	1.0%
8	Minimum Length of Lease Required	50 years plus length of loan
9	Usual Length of Loan	5–25 years
10	Investment	Yes
11	Refinance	Yes
12	Owner Occupier	Yes
13	Residential Investment	Yes
14	Development Finance	No
15	Residential Development	No
16	Hotels and Leisure	No
17	Healthcare	Yes
18	Bridging and/or Mezzanine Finance	No

Deutsche Bank AG

Asset Finance
1 Great Winchester Street
London
EC2N 2DB

Tel 020 7545 8000
Fax 020 7545 7130
www.deutsche-bank.de
london@dzbank.de

Contact
Vaughan Williams
020 7545 1652
vaughan.williams@db.com

Pamela Smith
020 7545 7520
pamela.smith@db.com

Criteria

1	Maximum Loan to Value	100%
2	Benchmark Lending Rate	3 month Libor
3	Margin Spread	Varies
4	Fixed	Yes
5	Minimum Loan	£25,000,000
6	Maximum Loan	None
7	Arrangement Fee Spread	Varies
8	Minimum Length of Lease Required	Varies
9	Usual Length of Loan	1 month–25 years
10	Investment	Yes
11	Refinance	Yes
12	Owner Occupier	No
13	Residential Investment	No
14	Development Finance	Yes, pre-let
15	Residential Development	No
16	Hotels and Leisure	No
17	Healthcare	No
18	Bridging and/or Mezzanine Finance	Yes

Remarks

Deutsche Bank bases lending decisions on the calibre of the borrower, the quality of the covenant, the property and the cash flow.

Deutsche Genossenschafts Hypothekenbank AG

Rosen Strasse 2
200095 Hamburg
Germany

10 Aldersgate Street
London
EC1A 4HJ

Tel 00 49 (0) 40 33 340
(Head Office)
020 7776 7610
(London Representative Office)
Fax 020 7776 7619
(London Representative Office)
www.dghyp.de

Contact

Andrew Goodbody
Head of London Representative Of
andrew.goodbody@dghyp.de

Criteria

1	Maximum Loan to Value	80%
2	Benchmark Lending Rate	3 month Libor
3	Margin Spread	0.5–2.0%
4	Fixed	Yes
5	Minimum Loan	£10,000,000
6	Maximum Loan	None
7	Arrangement Fee Spread	0.2–0.75%
8	Minimum Length of Lease Required	Freehold only
9	Usual Length of Loan	5–20 years
10	Investment	Yes
11	Refinance	Yes
12	Owner Occupier	No
13	Residential Investment	No
14	Development Finance	Partly pre-let only
15	Residential Development	No
16	Hotels and Leisure	Hotels — Yes Leisure — No
17	Healthcare	No
18	Bridging and/or Mezzanine Finance	No

Deutsche Hypo

City Tower – Level 19
40 Basinghall Street
London
EC2V 5DE

Tel 020 7920 0100
Fax 020 7920 0110
www.deutschehypo.de

Contact
Walter Love

Criteria

1	Maximum Loan to Value	75%
2	Benchmark Lending Rate	Libor
3	Margin Spread	0.80–1.20%
4	Fixed	Yes
5	Minimum Loan	£10,000,000
6	Maximum Loan	£125,000,000
7	Arrangement Fee Spread	0.2–0.5%
8	Minimum Length of Lease Required	5 years
9	Usual Length of Loan	3–10 years
10	Investment	Yes
11	Refinance	Yes
12	Owner Occupier	Yes
13	Residential Investment	Pre-let and pre-sold only
14	Development Finance	Yes
15	Residential Development	Yes
16	Hotels and Leisure	No
17	Healthcare	No
18	Bridging and/or Mezzanine Finance	No

Deutsche Postbank AG

61 Queen Street
London
EC4R 1AF

Tel 020 7634 2393
Fax 020 7248 4301
www.postbank.com

Contact
Gareth Jones
Real Estate Finance Director
gareth.jones@deutschepostbank.cc

John Cole
Real Estate Finance Manager
john.cole@deutschepostbank.co.uk

Suzanne Wadsworth
Real Estate Finance Manager
suzanne.wadsworth@deutschepostbank.c

Criteria

1	Maximum Loan to Value	90%
2	Benchmark Lending Rate	Libor
3	Margin Spread	0.5–1.0%
4	Fixed	Yes
5	Minimum Loan	£1,000,000
6	Maximum Loan	None
7	Arrangement Fee Spread	0.5–1.0%
8	Minimum Length of Lease Required	Varies
9	Usual Length of Loan	3–10 years
10	Investment	Yes
11	Refinance	Yes
12	Owner Occupier	Yes
13	Residential Investment	No
14	Development Finance	Yes
15	Residential Development	No
16	Hotels and Leisure	Yes
17	Healthcare	Yes
18	Bridging and/or Mezzanine Finance	Yes

Dudley Building Society

Dudley House
Stone Street
Dudley
DY1 1NP

Tel 01384 231414
Fax 01384 233250
www.dudleybuildingsociety.co.uk

Contact

John Robb
Mortgage Manager – Existing
Mortgage Accounts
01384 216834
mortgageadmin@dudleybuildingsociety.co

Samantha Hulse
Mortgage Sales
01384 216263
mortgagesales@dudleybuildingsociety.co.u

Criteria

1	Maximum Loan to Value	85%
2	Benchmark Lending Rate	Base
3	Margin Spread	1.0–3.0%
4	Fixed	Yes
5	Minimum Loan	£35,000
6	Maximum Loan	£1,000,000
7	Arrangement Fee Spread	1.0–3.00%
8	Minimum Length of Lease Required	40 years
9	Usual Length of Loan	Up to 25 years
10	Investment	Yes
11	Refinance	Yes
12	Owner Occupier	Yes
13	Residential Investment	Yes
14	Development Finance	No
15	Residential Development	No
16	Hotels and Leisure	No
17	Healthcare	No
18	Bridging and/or Mezzanine Finance	No

Dunbar Bank

9-15 Sackville Street
London
W1A 2JP

Tel 020 7437 7844
Fax 020 7437 3953
www.dunbarbank.co.uk

Contact

Paul Keay
Loans Director
020 7437 7844

Nigel Jackson
Loans Director
020 7437 7844

Steve Dry
Lending Manager
020 7437 7844

Criteria

1	Maximum Loan to Value	80%
2	Benchmark Lending Rate	Dunbar Bank base rate/ 3 month Libor
3	Margin Spread	2.0–4.0%
4	Fixed	Yes
5	Minimum Loan	£250,000
6	Maximum Loan	£20,000,000
7	Arrangement Fee Spread	1.0%
8	Minimum Length of Lease Required	50 years
9	Usual Length of Loan	Up to 5 years
10	Investment	Yes
11	Refinance	Yes
12	Owner Occupier	No
13	Residential Investment	Yes
14	Development Finance	Yes
15	Residential Development	Yes
16	Hotels and Leisure	No
17	Healthcare	No
18	Bridging and/or Mezzanine Finance	No

Remarks

Regional offices in Birmingham, Manchester, Leeds, Newcastle and Glasgow.

Dunfermline Building Society

Caledonia House
Carnegie Avenue
Dunfermline
KY11 8PJ

Tel 01383 627727
Fax 01383 627800
www.dunfermline.com

Contact
George Blyth
Manager of Consumer Lending

Criteria

1	Maximum Loan to Value	75%
2	Benchmark Lending Rate	Building society cost of funds
3	Margin Spread	1.0–3.0%
4	Fixed	Yes
5	Minimum Loan	£2,000
6	Maximum Loan	£250,000
7	Arrangement Fee Spread	1.0%
8	Minimum Length of Lease Required	30 years plus length of loan
9	Usual Length of Loan	5–20 years
10	Investment	Yes
11	Refinance	Yes
12	Owner Occupier	Yes
13	Residential Investment	Yes
14	Development Finance	Yes
15	Residential Development	Yes
16	Hotels and Leisure	Yes
17	Healthcare	Yes
18	Bridging and/or Mezzanine Finance	No

Remarks

Regional lender specialising in Scotland and the North of England.

DZ Bank

10 Aldersgate Street
London
EC1A 4HJ

Tel 020 7776 6000
Fax 020 7776 6100
www.dzbank.de

Contact
Holger Wessling
General Manager

Helen Jones
Loan Syndication
020 7776 6228
helen.jones@dzbank.de

Criteria

1	Maximum Loan to Value	90%
2	Benchmark Lending Rate	Base
3	Margin Spread	Varies
4	Fixed	Yes
5	Minimum Loan	£2,500,000
6	Maximum Loan	None
7	Arrangement Fee Spread	0.25–0.75%
8	Minimum Length of Lease Required	Varies
9	Usual Length of Loan	1–7 years
10	Investment	Yes
11	Refinance	Yes
12	Owner Occupier	No
13	Residential Investment	No
14	Development Finance	Yes
15	Residential Development	No
16	Hotels and Leisure	Yes
17	Healthcare	Yes
18	Bridging and/or Mezzanine Finance	Yes

Eagle Star Mortgages Ltd

Eagle Star Centre
Montpellier Drive
Cheltenham
Glos
GL53 7LQ

Tel 01242 221311
Fax 01242 221554

Contact
Brian Cosgrove

Criteria

1	Maximum Loan to Value	85%
2	Benchmark Lending Rate	Libor
3	Margin Spread	0.75–1.25%
4	Fixed	Yes
5	Minimum Loan	£500,000
6	Maximum Loan	£5,000,000
7	Arrangement Fee Spread	1.0%
8	Minimum Length of Lease Required	60 years
9	Usual Length of Loan	5–25 years
10	Investment	Yes
11	Refinance	Yes
12	Owner Occupier	Yes
13	Residential Investment	No
14	Development Finance	No
15	Residential Development	No
16	Hotels and Leisure	No
17	Healthcare	No
18	Bridging and/or Mezzanine Finance	No

Early Shilton Building Society

22 The Hollow
Early Shilton
Leicester
LE9 7NB

Tel 01455 844422
Fax 01455 845857
www.esbs.co.uk

Contact
Mortgage Team
0845 051 4422
mortgage@esbc.co.uk

Criteria

1	Maximum Loan to Value	80%
2	Benchmark Lending Rate	Base
3	Margin Spread	1.0–2.0%
4	Fixed	Yes
5	Minimum Loan	£50,000
6	Maximum Loan	£500,000
7	Arrangement Fee Spread	1.0–2.0%
8	Minimum Length of Lease Required	50 years
9	Usual Length of Loan	5–25 years
10	Investment	Yes
11	Refinance	Yes
12	Owner Occupier	Yes
13	Residential Investment	Yes
14	Development Finance	No
15	Residential Development	No
16	Hotels and Leisure	No
17	Healthcare	No
18	Bridging and/or Mezzanine Finance	No

Ecology Building Society

7 Belton Road
Silsden
Keighley
West Yorkshire
BD20 0EE

Tel 0845 674 5566
Fax 01535 650780

Contact
Pam Waring
General Manager
01535 650770
loans@ecology.co.uk

Criteria

1	Maximum Loan to Value	75%
2	Benchmark Lending Rate	Base
3	Margin Spread	0.5–1.0%
4	Fixed	Yes
5	Minimum Loan	£25,000
6	Maximum Loan	£750,000
7	Arrangement Fee Spread	1.0%
8	Minimum Length of Lease Required	75 years
9	Usual Length of Loan	5–15 years
10	Investment	Yes
11	Refinance	Yes
12	Owner Occupier	Yes
13	Residential Investment	No
14	Development Finance	No
15	Residential Development	No
16	Hotels and Leisure	No
17	Healthcare	No
18	Bridging and/or Mezzanine Finance	No

Egg Mortgages

Tel 08456 000 290
www.new.egg.com

Criteria

1	Maximum Loan to Value	100%
2	Benchmark Lending Rate	Libor
3	Margin Spread	1.0–2.0%
4	Fixed	Yes
5	Minimum Loan	£12,500
6	Maximum Loan	None
7	Arrangement Fee Spread	0.5–1.0%
8	Minimum Length of Lease Required	Varies
9	Usual Length of Loan	3–25 years
10	Investment	Yes
11	Refinance	Yes
12	Owner Occupier	Yes
13	Residential Investment	Yes
14	Development Finance	Yes
15	Residential Development	Yes
16	Hotels and Leisure	Yes
17	Healthcare	Yes
18	Bridging and/or Mezzanine Finance	Yes

Erste Bank

68 Cornhill
London
EC3V 3QE

Tel 020 7621 5000
Fax 020 7283 7142

Contact
Richard Mansfield
Head of Commercial Propery
020 7621 5018
rmm@erstebank.com

Criteria

1	Maximum Loan to Value	90% for senior debt 90–95% for senior mezzanine debt
2	Benchmark Lending Rate	Libor
3	Margin Spread	1.0–1.75%
4	Fixed	Yes
5	Minimum Loan	£5,000,000 (for senior debt)
6	Maximum Loan	£25,000,000 (for senior debt)
7	Arrangement Fee Spread	Varies
8	Minimum Length of Lease Required	None
9	Usual Length of Loan	5–10 years
10	Investment	Yes
11	Refinance	Yes
12	Owner Occupier	No
13	Residential Investment	No
14	Development Finance	Yes, pre-let and part pre-let
15	Residential Development	No
16	Hotels and Leisure	No
17	Healthcare	No
18	Bridging and/or Mezzanine Finance	Yes (see remarks)

Remarks

Pre-let and part pre-let development considered for parties with strong track records and business plan. Providers of mezzanine debt (as part of combined senior/mezzanine debt facility).

Eurohypo

4th Floor
90 Long Acre
WC2E 9RA

www.eurohypo.com

Contact
Maxim Sinclair
Head of London Branch
020 7759 7645
020 7759 7671

Mark Titcomb
Head of Origination
020 7759 7607
020 7759 7672

Criteria

1	Maximum Loan to Value	90%
2	Benchmark Lending Rate	Cost of funds
3	Margin Spread	Typically 0.85–1.25%
4	Fixed	Yes
5	Minimum Loan	£10,000,000 investment £20,000,000 development
6	Maximum Loan	None
7	Arrangement Fee Spread	0.35–1.00%
8	Minimum Length of Lease Required	60 years
9	Usual Length of Loan	5–15 years
10	Investment	Yes
11	Refinance	Yes
12	Owner Occupier	Yes
13	Residential Investment	Yes
14	Development Finance	Yes
15	Residential Development	Yes
16	Hotels and Leisure	Yes
17	Healthcare	Yes
18	Bridging and/or Mezzanine Finance	Yes

First National Bank

College Road
Harrow
Middx
HA1 1FB

Tel 020 8861 1313
Fax 020 8861 4540

Contact
Jeff Watson

A Turzynski

Criteria

1	Maximum Loan to Value	90%
2	Benchmark Lending Rate	Finance house base rate 3 month Libor
3	Margin Spread	1.5–4.0%
4	Fixed	No
5	Minimum Loan	£50,000
6	Maximum Loan	£2,000,000
7	Arrangement Fee Spread	Varies
8	Minimum Length of Lease Required	40 years
9	Usual Length of Loan	Up to 25 years
10	Investment	Yes
11	Refinance	Ys
12	Owner Occupier	Yes
13	Residential Investment	Yes
14	Development Finance	No
15	Residential Development	No
16	Hotels and Leisure	Yes
17	Healthcare	No
18	Bridging and/or Mezzanine Finance	No

First National Building Society

Sketan House
Booterstown
Co Dublin
Ireland

Tel 00 3531 288 5211
Fax 00 3531 288 5300

Contact
Commercial Mortgage Manager
at nearest branch

Criteria

1	Maximum Loan to Value	80%
2	Benchmark Lending Rate	Standard variable rate
3	Margin Spread	Up to 2.5%
4	Fixed	Yes
5	Minimum Loan	£15,001
6	Maximum Loan	£1,000,000
7	Arrangement Fee Spread	1.0%
8	Minimum Length of Lease Required	75 years
9	Usual Length of Loan	10–30 years
10	Investment	Yes
11	Refinance	Yes
12	Owner Occupier	Yes
13	Residential Investment	Yes
14	Development Finance	No
15	Residential Development	Yes
16	Hotels and Leisure	Yes
17	Healthcare	Yes
18	Bridging and/or Mezzanine Finance	No

Remarks

Irish Republic only.

Furness Building Society

51-55 Duke Street
Barrow-in-Furness
Cumbria
LA14 1RT

Tel 01229 824560
Fax 01229 837043
www.furnessbs.co.uk

Contact
Phillip Blake
Mortgage Controller

Criteria

1	Maximum Loan to Value	75%
2	Benchmark Lending Rate	Cost of funds
3	Margin Spread	1.0–2.0%
4	Fixed	Yes
5	Minimum Loan	None
6	Maximum Loan	£250,000
7	Arrangement Fee Spread	1.0%
8	Minimum Length of Lease Required	N/A
9	Usual Length of Loan	5–25 years
10	Investment	Yes
11	Refinance	Yes
12	Owner Occupier	Yes
13	Residential Investment	Yes
14	Development Finance	Pre-let only
15	Residential Development	No
16	Hotels and Leisure	Yes
17	Healthcare	No
18	Bridging and/or Mezzanine Finance	No

Remarks

Within 100-mile radius of Barrow-in-Furness.

GE Real Estate

30 Berkeley Square
London
W1J 6EW

Tel 020 7302 6000
Fax 020 7302 6830
www.gerealestate.co.uk

Contact
Neil Harris
Head of Asset Management
020 7302 6053
neil.harris@ge.com

Criteria

1	Maximum Loan to Value	90%
2	Benchmark Lending Rate	Libor
3	Margin Spread	Depends
4	Fixed	Yes
5	Minimum Loan	£50,000,000
6	Maximum Loan	None
7	Arrangement Fee Spread	Depends
8	Minimum Length of Lease Required	N/A
9	Usual Length of Loan	5 years
10	Investment	Yes
11	Refinance	Yes
12	Owner Occupier	No
13	Residential Investment	Yes
14	Development Finance	Yes
15	Residential Development	No
16	Hotels and Leisure	Yes
17	Healthcare	Yes
18	Bridging and/or Mezzanine Finance	Mezzanine finance

GMAC-RFC

Eastern Gate
Brants Bridge
Bracknell
Berkshire
RG12 9BZ

Tel 01344 478478
Fax 01344 478050
www.gmacrfc.co.uk

Contact
Commercial Mortgage Manager
at nearest branch

Criteria

1	Maximum Loan to Value	90%
2	Benchmark Lending Rate	Libor
3	Margin Spread	Varies
4	Fixed	Yes
5	Minimum Loan	£1,000,000
6	Maximum Loan	£10,000,000
7	Arrangement Fee Spread	0.50–0.75%
8	Minimum Length of Lease Required	Varies
9	Usual Length of Loan	1–5 years
10	Investment	Yes
11	Refinance	Yes
12	Owner Occupier	No
13	Residential Investment	No
14	Development Finance	Yes
15	Residential Development	Yes
16	Hotels and Leisure	Yes
17	Healthcare	Yes
18	Bridging and/or Mezzanine Finance	Yes

Goldman Sachs Real Estate

Peterborough Court
133 Fleet Street
London
EC4A 2BB

Tel 020 7774 1000
www2.goldmansachs.com

Contact
James Fel Favero
Richard Georgi

Criteria

1	Maximum Loan to Value	100% subject to status
2	Benchmark Lending Rate	Base
3	Margin Spread	Varies
4	Fixed	No
5	Minimum Loan	£30,000,000
6	Maximum Loan	None
7	Arrangement Fee Spread	Varies
8	Minimum Length of Lease Required	None
9	Usual Length of Loan	Varies
10	Investment	Yes
11	Refinance	Yes
12	Owner Occupier	Yes
13	Residential Investment	No
14	Development Finance	Yes
15	Residential Development	No
16	Hotels and Leisure	Yes
17	Healthcare	No
18	Bridging and/or Mezzanine Finance	Yes

Remarks

Provides funding for major property investments and developments to recognised and established companies.

Hampshire Trust Plc

Fareham House
69 High Street
Fareham
Hampshire
PO16 7BB

Tel 01329 234294
Fax 01329 285910
www.hampshire-trust.co.uk

Contact
Simon Atherton
Associate Director
sa@htplc.co.uk

Berry Jones
Property Underwriter
bj@htplc.co.uk

Christine Rickeard
Loans Administrator
cr@htplc.co.uk

Criteria

1	Maximum Loan to Value	75%
2	Benchmark Lending Rate	Libor
3	Margin Spread	0–3.5% over Average 2.5%
4	Fixed	Not available
5	Minimum Loan	£50,000
6	Maximum Loan	£1,000,000
7	Arrangement Fee Spread	1.0–1.5%
8	Minimum Length of Lease Required	50 years plus loan term
9	Usual Length of Loan	3 months–20 years
10	Investment	Yes
11	Refinance	Yes
12	Owner Occupier	Yes
13	Residential Investment	Yes
14	Development Finance	Yes
15	Residential Development	Yes
16	Hotels and Leisure	Yes
17	Healthcare	Yes
18	Bridging and/or Mezzanine Finance	Yes

Remarks

All applications are individually assessed without recourse to credit scoring or the like. Prompt decisions given. Adverse credit and limited accounting information are viewed sympathetically. A high-speed bridging loan facility is available.

Heritable Bank Ltd

8 Hill Street
Berkeley Square
London
W1J 5NG

Tel 020 7493 6621
Fax 020 7629 1958
www.heritable.co.uk

Contact
Nick Barrett
MD Structured Property Finance
corporate@heritable.co.uk

Criteria

1	Maximum Loan to Value	85%
2	Benchmark Lending Rate	Base
3	Margin Spread	2.5–4.0%
4	Fixed	Yes
5	Minimum Loan	£1,000,000
6	Maximum Loan	None
7	Arrangement Fee Spread	0.5–1.0%
8	Minimum Length of Lease Required	None
9	Usual Length of Loan	Up to 36 months
10	Investment	No
11	Refinance	Yes
12	Owner Occupier	No
13	Residential Investment	Yes
14	Development Finance	Yes — residential and commercial
15	Residential Development	Yes
16	Hotels and Leisure	No
17	Healthcare	No
18	Bridging and/or Mezzanine Finance	Yes

Remarks

The bank lends to all types of residential development throughout England.

Hinckley & Rugby Building Society

81 Upper Bond Street
Hinckley
Leicestershire
LE10 1DG

Tel 01455 251234
Fax 01455 636237
www.hrbs.co.uk

Contact
Martyn Shortland
Mortgage Service Manager
m.shortland@hrbs.co.uk

Julie Chapman
Assistant Mortgage Service Manage
j.chapman@hrbs.co.uk

Criteria

1	Maximum Loan to Value	75%
2	Benchmark Lending Rate	Base mortgage rate + 1.5%
3	Margin Spread	1.5–2.0%
4	Fixed	No
5	Minimum Loan	£25,000
6	Maximum Loan	£500,000
7	Arrangement Fee Spread	None
8	Minimum Length of Lease Required	75 years from date of advance
9	Usual Length of Loan	Up to 15 years
10	Investment	Yes
11	Refinance	Possible
12	Owner Occupier	Yes
13	Residential Investment	Yes
14	Development Finance	No
15	Residential Development	No
16	Hotels and Leisure	No
17	Healthcare	No
18	Bridging and/or Mezzanine Finance	No

Remarks

Commercial lending restricted to society's core business area in the East Midlands.

Julian Hodge Bank Ltd

10 Windsor Place
Cardiff
CF10 3BX

Tel 029 2080 3095
Fax 029 2034 4061
www.julianhodgebank.com

Contact
Damon Lawrence
damon.lawrence@jhb.co.uk

Paul Budd
paul.budd@jhb.co.uk

Criteria

1	Maximum Loan to Value	75% for 1st deal
2	Benchmark Lending Rate	Bank base
3	Margin Spread	1.75–3.0%
4	Fixed	Yes
5	Minimum Loan	£500,000
6	Maximum Loan	£25,000,000
7	Arrangement Fee Spread	Minimum 1.0%
8	Minimum Length of Lease Required	Negotiable
9	Usual Length of Loan	No minimum — 20 years
10	Investment	Yes
11	Refinance	Yes
12	Owner Occupier	Yes
13	Residential Investment	No
14	Development Finance	Yes
15	Residential Development	Yes
16	Hotels and Leisure	No
17	Healthcare	Yes
18	Bridging and/or Mezzanine Finance	Yes — negotiable

Holborn Securities Plc

3 Park Road
Regents Park
London
NW1 6AS

Tel 020 7616 6616
Fax 020 7616 6610

Contact
Henry Ejdelbaum

Glenn Roberts

Criteria

1	Maximum Loan to Value	85%
2	Benchmark Lending Rate	Libor
3	Margin Spread	1.5–3.5%
4	Fixed	No
5	Minimum Loan	£75,000
6	Maximum Loan	£1,000,000
7	Arrangement Fee Spread	0.5–1.25%
8	Minimum Length of Lease Required	Varies
9	Usual Length of Loan	1–20 years
10	Investment	Yes
11	Refinance	Yes
12	Owner Occupier	Yes
13	Residential Investment	Yes
14	Development Finance	Yes
15	Residential Development	Yes
16	Hotels and Leisure	Yes
17	Healthcare	Yes
18	Bridging and/or Mezzanine Finance	Yes

Remarks

Prefers bridging loans and enablement finance.

Holmesdale Building Society

43 Church Street
Reigate
Surrey
RH2 0AE

Tel 01737 245716
Fax 01737 246962
www.holmesdale.org.uk

Contact
Peter Bell
Mortgage Manager

Criteria

1	Maximum Loan to Value	80%
2	Benchmark Lending Rate	Base
3	Margin Spread	1.0–2.0%
4	Fixed	No
5	Minimum Loan	£50,000
6	Maximum Loan	£1,000,000
7	Arrangement Fee Spread	1.0–2.0%
8	Minimum Length of Lease Required	40 years
9	Usual Length of Loan	Up to 25 years
10	Investment	No
11	Refinance	Yes
12	Owner Occupier	Yes
13	Residential Investment	Yes
14	Development Finance	No
15	Residential Development	No
16	Hotels and Leisure	No
17	Healthcare	No
18	Bridging and/or Mezzanine Finance	No

HSBC Bank Plc

1st Floor
60 Queen Victoria Street
London
EC4N 4TR

Tel 08455 847732
www.hsbc.co.uk

Contact
Stuart Grant
Head of Corporate and
Structured Banking

Colin Stratton
Real Estate Specialist
08455 847721
colinstratton@hsbc.com

Mike O'Connor
Head of Corporate Banking, Hotels
020 7991 2945

Graham Harman
Real Estate Specialist
08455 845554
grahamharman@hsbc.com

Criteria

1	Maximum Loan to Value	90%
2	Benchmark Lending Rate	Interbank/base
3	Margin Spread	0.8–2.0%
4	Fixed	Yes
5	Minimum Loan	£10,000,000
6	Maximum Loan	None
7	Arrangement Fee Spread	0.5–1.0%
8	Minimum Length of Lease Required	None
9	Usual Length of Loan	5–7 years
10	Investment	Yes
11	Refinance	Yes
12	Owner Occupier	Yes
13	Residential Investment	Yes
14	Development Finance	Yes
15	Residential Development	Yes
16	Hotels and Leisure	Yes
17	Healthcare	No
18	Bridging and/or Mezzanine Finance	Yes

HSH Nordbank AG

Moorgate Hall
155 Moorgate
London
EC2M 6UJ

Tel 020 7972 9292
Fax 020 7972 9290
www.hsh-nordbank.co.uk

Contact
Ray McColl
Head of Real Estate
020 7903 6428
ray.mccoll@hsh-nordbank.co.uk

John Crocker
020 7903 7758
john.crocker@hsh-nordbank.co.uk

Hans Tamm
Loan Syndications
020 7903 7797
hans.tamm@hsh-nordbank.co.uk

Criteria

1	Maximum Loan to Value	90%
2	Benchmark Lending Rate	Libor
3	Margin Spread	0.5–1.0%
4	Fixed	Yes
5	Minimum Loan	£1,000,000
6	Maximum Loan	None
7	Arrangement Fee Spread	1%
8	Minimum Length of Lease Required	35 years
9	Usual Length of Loan	3–15 years
10	Investment	Yes
11	Refinance	Yes
12	Owner Occupier	No
13	Residential Investment	No
14	Development Finance	No
15	Residential Development	Yes
16	Hotels and Leisure	Yes
17	Healthcare	Yes
18	Bridging and/or Mezzanine Finance	No

Hypo Real Estate Bank International

21st Floor
30 St Mary Axe
London
EC3A 8BF

Tel 020 7743 7743
Fax 020 7743 7700
www.hyporealestate.com
london@hypointernational.com

Contact
Harin Thaker
CEO

Jairaj Amin
Managing Director

Simon Fisk
Managing Director

Walter Hampel
SD

Criteria

1	Maximum Loan to Value	100%
2	Benchmark Lending Rate	Libor
3	Margin Spread	0.5–1.0%
4	Fixed	Yes
5	Minimum Loan	£1,000,000
6	Maximum Loan	None
7	Arrangement Fee Spread	0.5–1.0%
8	Minimum Length of Lease Required	50 years
9	Usual Length of Loan	5–25 years
10	Investment	Yes
11	Refinance	Yes
12	Owner Occupier	Yes
13	Residential Investment	Yes
14	Development Finance	Yes
15	Residential Development	Yes
16	Hotels and Leisure	Yes
17	Healthcare	Yes
18	Bridging and/or Mezzanine Finance	Yes

Remarks

Hypo Real Estate Bank International is a mid-to-large ticket lender specialising in Commercial Real Estate Finance. The Hypo Real Estate Group is a global lender with offices in 25 countries globally.

Hypothekenbank Essen AG

6th Floor
60 Gracechurch Streer
London
EC3V 1HR

Tel 020 7283 3142
Fax 020 72 83 26 49
www.essenhyp.com

Contact
Rainer Polenz
General Manager
r.polenz@essenhyp.com

Criteria

1	Maximum Loan to Value	85%
2	Benchmark Lending Rate	Libor
3	Margin Spread	0.25–0.75%
4	Fixed	Yes
5	Minimum Loan	£5,000,000
6	Maximum Loan	None
7	Arrangement Fee Spread	0.25–0.75%
8	Minimum Length of Lease Required	75 years
9	Usual Length of Loan	5–15 years
10	Investment	Yes
11	Refinance	Yes
12	Owner Occupier	No
13	Residential Investment	No
14	Development Finance	Yes
15	Residential Development	Yes
16	Hotels and Leisure	Yes
17	Healthcare	Yes
18	Bridging and/or Mezzanine Finance	Yes

Intermediate Capital Group Plc

20 Old Broad Street
London
EC2N 1DP

Tel 020 7628 9898
Fax 020 7628 2268
www.icgplc.com

Contact
Tom Bartlam

Philip Keller
Finance, Managing Director

James Allen
Mezzanine Investment

Criteria

1	Maximum Loan to Value	Mezzanine finance and equity finance only
2	Benchmark Lending Rate	Base
3	Margin Spread	Varies
4	Fixed	No
5	Minimum Loan	£7,500,000
6	Maximum Loan	None
7	Arrangement Fee Spread	Varies
8	Minimum Length of Lease Required	None
9	Usual Length of Loan	Short-term funding
10	Investment	No
11	Refinance	No
12	Owner Occupier	No
13	Residential Investment	No
14	Development Finance	Yes, but only leisure
15	Residential Development	No
16	Hotels and Leisure	Yes
17	Healthcare	Yes
18	Bridging and/or Mezzanine Finance	Yes

Remarks

Providers of mezzanine and equity finance as well as immediate capital for buyouts. Primarily active in leisure, hotels and nursing homes. No pure property.

Investec Bank (UK) Ltd

2 Gresham Street
London
EC2V 7QP

Tel 020 7597 4010
Fax 020 7597 4070
www.investec.com/UnitedKingdom

Contact
Steven Cook

Gary Dobson

Jill Hagland

Peter Harris

John Pollington

Andrew Scotting

Criteria

1	Maximum Loan to Value	90%
2	Benchmark Lending Rate	3 month Libor/base rate
3	Margin Spread	1.5–3.0%
4	Fixed	Yes, max 5 years
5	Minimum Loan	£500,000
6	Maximum Loan	None
7	Arrangement Fee Spread	Negotiable
8	Minimum Length of Lease Required	Varies (short leases considered)
9	Usual Length of Loan	3 months–15 years
10	Investment	Yes
11	Refinance	Yes
12	Owner Occupier	No
13	Residential Investment	Yes
14	Development Finance	London and M25 vicinity. Other prime England locations considered.
15	Residential Development	Yes
16	Hotels and Leisure	No
17	Healthcare	No
18	Bridging and/or Mezzanine Finance	Yes

Remarks

In the commercial investment sector, only strong covenants or prime locations considered. Mezzanine funding considered on residential development up to 90% for "exit fee" arrangements. Offshore structures acceptable.

Ipswich Building Society

PO Box 547
Ispwich
Suffolk
IP3 9WZ

Tel 0845 2308686
www.ibs.co.uk

Contact
Ian Dockley
Financial Services Manager
iandockley@ibs.co.uk

Criteria

1	Maximum Loan to Value	80%
2	Benchmark Lending Rate	Base
3	Margin Spread	1.0–2.0%
4	Fixed	No
5	Minimum Loan	£100,000
6	Maximum Loan	£750,000
7	Arrangement Fee Spread	1.0–2.0%
8	Minimum Length of Lease Required	60 years
9	Usual Length of Loan	5–30 years
10	Investment	Yes
11	Refinance	Yes
12	Owner Occupier	Yes
13	Residential Investment	Yes
14	Development Finance	No
15	Residential Development	No
16	Hotels and Leisure	No
17	Healthcare	No
18	Bridging and/or Mezzanine Finance	No

Irish Permanent International

12/14 Ridgeway Street
Douglas
Isle of Man
IM1 1EN

Tel 01624 641641
Fax 01624 676795
www.irishpermanentintl.com
info@irishpermanentintl.com

Contact

Kevin Farrington
Lending Manager
01624 641607

Criteria

1	Maximum Loan to Value	80%
2	Benchmark Lending Rate	Base/Libor +
3	Margin Spread	1.25–2.0%
4	Fixed	On request
5	Minimum Loan	£350,000
6	Maximum Loan	None
7	Arrangement Fee Spread	0.5–1.0%
8	Minimum Length of Lease Required	5 years
9	Usual Length of Loan	5 years
10	Investment	Yes
11	Refinance	Yes
12	Owner Occupier	Yes — professional practices
13	Residential Investment	Yes
14	Development Finance	No
15	Residential Development	No
16	Hotels and Leisure	No
17	Healthcare	No
18	Bridging and/or Mezzanine Finance	No

KBC Bank NV

111 Old Broad Street
London
EC2N 1BR

Tel 020 7638 5812
Fax 020 7663 0151
www.kbc.com

Contact
Stephen Picton-Pegg
Head of Real Estate
020 7256 4866
stephen.picton-pegg@kbc.be

Robert Baskeyfield
Manager Real Estate
020 7256 4849
robert.baskeyfield@kbc.be

David Woodgate
Real Estate
020 7256 4826
david.woodgate@kbc.be

Criteria

1	Maximum Loan to Value	80%
2	Benchmark Lending Rate	Libor
3	Margin Spread	1.0–2.5%
4	Fixed	Yes
5	Minimum Loan	£3,000,000
6	Maximum Loan	£30,000,000
7	Arrangement Fee Spread	0.5–1.0%
8	Minimum Length of Lease Required	N/A
9	Usual Length of Loan	5–10 years
10	Investment	Yes
11	Refinance	Yes
12	Owner Occupier	Yes
13	Residential Investment	No
14	Development Finance	Yes
15	Residential Development	No
16	Hotels and Leisure	No
17	Healthcare	No
18	Bridging and/or Mezzanine Finance	No

Remarks

Investment property finance. Will consider both pre-let and speculative development for qualified borrowers (ie, with a proven track record and business plan). Residential development will be considered when part of a mixed use scheme.

Kensington Mortgages

Reading International Business Park
Basingstoke Road
Reading
RG2 6DB

Tel 0800 111 020
www.kmc.co.uk

Criteria

1	Maximum Loan to Value	80%
2	Benchmark Lending Rate	Base
3	Margin Spread	0.5–1.0%
4	Fixed	Yes
5	Minimum Loan	£50,000
6	Maximum Loan	£1,000,000
7	Arrangement Fee Spread	1%
8	Minimum Length of Lease Required	25 years
9	Usual Length of Loan	5–25 years
10	Investment	Yes
11	Refinance	Yes
12	Owner Occupier	Yes
13	Residential Investment	Yes
14	Development Finance	No
15	Residential Development	No
16	Hotels and Leisure	No
17	Healthcare	Yes
18	Bridging and/or Mezzanine Finance	Yes

Kent Reliance Building Society

Reliance House
Manor Road
Chatham
Kent
ME4 6AF

Tel 01634 848944
Fax 01634 830912
www.krbs.com

Contact
Adrian Wicken
Commercial Lending Manager

Criteria

1	Maximum Loan to Value	75% commercial 85% residential investment
2	Benchmark Lending Rate	Base
3	Margin Spread	1.5–3.0%
4	Fixed	Yes
5	Minimum Loan	£25,000
6	Maximum Loan	£5,000,000
7	Arrangement Fee Spread	0.5%, minimum £250
8	Minimum Length of Lease Required	Term on loan plus 50 years
9	Usual Length of Loan	3–25 years
10	Investment	Yes
11	Refinance	Yes
12	Owner Occupier	Yes
13	Residential Investment	Yes
14	Development Finance	Yes
15	Residential Development	Yes
16	Hotels and Leisure	Yes
17	Healthcare	Yes
18	Bridging and/or Mezzanine Finance	No

Landesbank Hessen-Thüringen (Helaba)

95 Queen Victoria Street
London
EC4V 4HN

Tel 020 7334 4500
Fax 020 7330 4502
www.helaba.de
propertyfinance.uk@helaba.de

Contact

Peter A B Wilson
Head of Real Estate
020 7334 4556
propertyfiance.uk@helaba.de

Graham Chapman

Chris Allsop

Laura Williams

Criteria

1	Maximum Loan to Value	80%
2	Benchmark Lending Rate	Libor
3	Margin Spread	0.75–1.75%
4	Fixed	10 years available
5	Minimum Loan	£10,000,000
6	Maximum Loan	None
7	Arrangement Fee Spread	0.35–0.80%
8	Minimum Length of Lease Required	None
9	Usual Length of Loan	3–10 years
10	Investment	Up to 10 years
11	Refinance	Yes
12	Owner Occupier	No
13	Residential Investment	Yes
14	Development Finance	Speculative: up to 60% of cost Pre-let: up to 90% of cost
15	Residential Development	Yes
16	Hotels and Leisure	No
17	Healthcare	No
18	Bridging and/or Mezzanine Finance	No

Remarks

Portfolio finance a speciality.

Leeds Building Society

105 Albion Street
Leeds
LS1 5AS

Tel 0113 225 7777
Fax 0113 225 7659
www.leedsbuildingsociety.co.uk

Contact
Linsey Herrington
Marketing Communications
Team Leader
l.herrington@leedsbuildingsociety.co.u

Criteria

1	Maximum Loan to Value	80%
2	Benchmark Lending Rate	Base
3	Margin Spread	1.0–3.0%
4	Fixed	Yes
5	Minimum Loan	£35,000
6	Maximum Loan	£1,000,000
7	Arrangement Fee Spread	1.0–3.0%
8	Minimum Length of Lease Required	40 years
9	Usual Length of Loan	5–25 years
10	Investment	Yes
11	Refinance	Yes
12	Owner Occupier	Yes
13	Residential Investment	Yes
14	Development Finance	No
15	Residential Development	No
16	Hotels and Leisure	No
17	Healthcare	No
18	Bridging and/or Mezzanine Finance	No

Leek United Building Society

50 St Edward Street
Leek
Staffordshire
ST13 5DL

Tel 01538 384151
Fax 01538 399179
www.leekunited.co.uk

Contact
Ruth Taylor
Mortgage Applications Manager

Criteria

1	Maximum Loan to Value	75%
2	Benchmark Lending Rate	N/A
3	Margin Spread	1.0–2.5%
4	Fixed	Yes
5	Minimum Loan	£25,000
6	Maximum Loan	£500,000
7	Arrangement Fee Spread	1.0%
8	Minimum Length of Lease Required	N/A
9	Usual Length of Loan	5–20 years
10	Investment	Yes
11	Refinance	Yes
12	Owner Occupier	Yes
13	Residential Investment	Yes
14	Development Finance	No
15	Residential Development	Yes
16	Hotels and Leisure	No
17	Healthcare	No
18	Bridging and/or Mezzanine Finance	No

Legal & General Bank

Temple Court
11 Queen Victoria Street
London
EC4N 4TP

Tel 0800 096 6959

Contact
Peter Saders

Criteria

1	Maximum Loan to Value	80%
2	Benchmark Lending Rate	3 month Libor
3	Margin Spread	1.0–2.0%
4	Fixed	Yes, up to 30 years
5	Minimum Loan	£1,000,000
6	Maximum Loan	None
7	Arrangement Fee Spread	0.25–1.0%
8	Minimum Length of Lease Required	70 years
9	Usual Length of Loan	10–25 years
10	Investment	Yes
11	Refinance	Yes
12	Owner Occupier	Yes
13	Residential Investment	No
14	Development Finance	No
15	Residential Development	No
16	Hotels and Leisure	No
17	Healthcare	No
18	Bridging and/or Mezzanine Finance	No

Lehman Brothers

Europe and Middle East Headquarters
25 Bank Street
London
E14 5LE

Tel 020 7102 1000
www.lehman.com

Criteria

1	Maximum Loan to Value	100%
2	Benchmark Lending Rate	Libor
3	Margin Spread	0.5–1.0%
4	Fixed	Yes
5	Minimum Loan	£10,000,000
6	Maximum Loan	None
7	Arrangement Fee Spread	Negotiable
8	Minimum Length of Lease Required	Varies
9	Usual Length of Loan	1–10 years
10	Investment	Yes
11	Refinance	Yes
12	Owner Occupier	Yes
13	Residential Investment	No
14	Development Finance	Yes
15	Residential Development	Yes
16	Hotels and Leisure	Yes
17	Healthcare	Yes
18	Bridging and/or Mezzanine Finance	Yes

Liverpool & Victoria

County Gates House
300 Poole Road
Bournemouth
Dorset
BH1 3NQ

Tel 01202 295544
Fax 01202 558712
www.lv.com

Criteria

1	Maximum Loan to Value	80%
2	Benchmark Lending Rate	Libor
3	Margin Spread	Varies
4	Fixed	No
5	Minimum Loan	£50,000
6	Maximum Loan	£1,000,000
7	Arrangement Fee Spread	Varies
8	Minimum Length of Lease Required	None
9	Usual Length of Loan	1–10 years
10	Investment	Yes
11	Refinance	Yes
12	Owner Occupier	No
13	Residential Investment	No
14	Development Finance	No
15	Residential Development	No
16	Hotels and Leisure	No
17	Healthcare	No
18	Bridging and/or Mezzanine Finance	No

Lloyds TSB

PO Box 112
Canons House
Canons Way
Bristol
BS99 7LB

Tel 0800 783 3534
www.lloydstsb.com

Contact
J M Whitaker

John Spence

Criteria

1	Maximum Loan to Value	85%
2	Benchmark Lending Rate	Bank base/Libor/ cost of funds
3	Margin Spread	1.0–4.5%
4	Fixed	5–15 years
5	Minimum Loan	£1,000
6	Maximum Loan	£25,000,000
7	Arrangement Fee Spread	Negotiable
8	Minimum Length of Lease Required	Varies
9	Usual Length of Loan	1–25 years
10	Investment	Yes
11	Refinance	Yes
12	Owner Occupier	Yes
13	Residential Investment	Yes
14	Development Finance	Yes, pre-let
15	Residential Development	No
16	Hotels and Leisure	No
17	Healthcare	Yes
18	Bridging and/or Mezzanine Finance	No

London Mortgage Company

6 Broadgate
London
EC2M 2QS

Tel 020 7920 1123
Fax 020 7920 1123
www.l-m-c.co.uk

Contact

Sharon O'Callaghan
Director of Regional Sales
(London and the Home Counties)
07796 278922

Maria Betti
Senior Regional Sales Manager
(South and South East)
07919 328773

Stuart Day
Regional Sales Manager
(The Midlands and East Anglia)
07799 696035

Scott Bullen
Regional Sales Manager
(The North)

Criteria

1	Maximum Loan to Value	85%
2	Benchmark Lending Rate	Libor
3	Margin Spread	1.0–2.0%
4	Fixed	Yes
5	Minimum Loan	£75,000
6	Maximum Loan	£1,000,000
7	Arrangement Fee Spread	1.0–3.00%
8	Minimum Length of Lease Required	Varies
9	Usual Length of Loan	1–25 years
10	Investment	Yes
11	Refinance	Yes
12	Owner Occupier	Yes
13	Residential Investment	Yes
14	Development Finance	Yes
15	Residential Development	Yes
16	Hotels and Leisure	Yes
17	Healthcare	Yes
18	Bridging and/or Mezzanine Finance	Yes

Manchester Building Society

125 Portland Street
Manchester
M1 4QD

Tel 0161 833 8888
Fax 0161 833 8889
www.themanchester.co.uk

Contact
David Cowie

Criteria

1	Maximum Loan to Value	85%
2	Benchmark Lending Rate	Standard variable rate
3	Margin Spread	Varies
4	Fixed	Yes
5	Minimum Loan	£50,000
6	Maximum Loan	£1,000,000
7	Arrangement Fee Spread	0.5–1.5%
8	Minimum Length of Lease Required	40 years
9	Usual Length of Loan	5–30 years
10	Investment	Yes
11	Refinance	Yes
12	Owner Occupier	Yes
13	Residential Investment	Yes
14	Development Finance	No
15	Residential Development	Yes
16	Hotels and Leisure	Yes
17	Healthcare	Yes
18	Bridging and/or Mezzanine Finance	Yes

Remarks

Prefers to lend within 50-mile radius of Manchester.

W M Mann & Co (Investments) Ltd

201 Bath Street
Glasgow
G2 4H2

Tel 0141 248 4936
Fax 0141 221 2976

Contact
Bruce Mann
Director

Criteria

1	Maximum Loan to Value	70%
2	Benchmark Lending Rate	Clydesdale bank base
3	Margin Spread	3.75–4.0%
4	Fixed	No
5	Minimum Loan	£25,500
6	Maximum Loan	£900,000
7	Arrangement Fee Spread	1.0–2.0%
8	Minimum Length of Lease Required	Freehold only
9	Usual Length of Loan	6 months–15 years
10	Investment	Yes
11	Refinance	Occasionally
12	Owner Occupier	Yes
13	Residential Investment	Yes
14	Development Finance	Yes
15	Residential Development	Yes
16	Hotels and Leisure	Yes
17	Healthcare	Yes
18	Bridging and/or Mezzanine Finance	Yes

Remarks

Commercial lending only in Scotland, especially resindential and retail development.

Mansfield Building Society

Regent House
Regent Street
Mansfield
Notts
NG18 1SS

Tel 01623 676300
Fax 01623 420261
www.mansfieldbs.co.uk

Contact
Maggie Smith
Mortgage Executive

Criteria

1	Maximum Loan to Value	80%
2	Benchmark Lending Rate	Standard variable rate
3	Margin Spread	+1% subject to assessment of the proposition
4	Fixed	No
5	Minimum Loan	£25,001
6	Maximum Loan	£500,000
7	Arrangement Fee Spread	1.0%
8	Minimum Length of Lease Required	N/A
9	Usual Length of Loan	15 years
10	Investment	Yes
11	Refinance	Yes
12	Owner Occupier	Yes
13	Residential Investment	Yes
14	Development Finance	No
15	Residential Development	Yes
16	Hotels and Leisure	Yes
17	Healthcare	No
18	Bridging and/or Mezzanine Finance	No

Market Harborough Building Society

The Square
Market Harborough
Leicestershire
LE16 7PD

Tel 01858 412250
Fax 01858 412254
www.mhbs.co.uk

Contact
Paul Thilliy
Assistant General Manager
of Customer Services

Criteria

1	Maximum Loan to Value	80%
2	Benchmark Lending Rate	Base
3	Margin Spread	1.0–2.0%
4	Fixed	No
5	Minimum Loan	£25,000
6	Maximum Loan	£500,000
7	Arrangement Fee Spread	1.0–2.0%
8	Minimum Length of Lease Required	60 years
9	Usual Length of Loan	Up to 25 years
10	Investment	Yes
11	Refinance	Yes
12	Owner Occupier	Yes
13	Residential Investment	Yes
14	Development Finance	No
15	Residential Development	No
16	Hotels and Leisure	No
17	Healthcare	No
18	Bridging and/or Mezzanine Finance	No

Marsden Building Society

6-20 Russell Street
Nelson
Lancashire
BB9 7NJ

Tel 01282 440500
www.marsdenbs.co.uk

Contact
Steve Robinson
Assistant General Manager of Lend
s.robinson@marsden.co.uk

Criteria

1	Maximum Loan to Value	75%
2	Benchmark Lending Rate	Standard variable rate
3	Margin Spread	1.0–1.25%
4	Fixed	Yes
5	Minimum Loan	£15,000
6	Maximum Loan	£500,000
7	Arrangement Fee Spread	1.0%
8	Minimum Length of Lease Required	N/A
9	Usual Length of Loan	10–25 years
10	Investment	Yes
11	Refinance	Yes
12	Owner Occupier	Yes
13	Residential Investment	Yes
14	Development Finance	No
15	Residential Development	No
16	Hotels and Leisure	Yes
17	Healthcare	No
18	Bridging and/or Mezzanine Finance	No

Remarks

Regional lending only. 100-mile radius of Marsden.

Melton Mowbray Building Society

Leicester Road
Melton Mowbray
Leicestershire
LE13 0DB

Tel 01664 414141
Fax 01664 414040
www.mmbs.co.uk

Contact
Martin McDermott
Director of Opertions

Criteria

1	Maximum Loan to Value	80%
2	Benchmark Lending Rate	Base
3	Margin Spread	1.0–2.0%
4	Fixed	No
5	Minimum Loan	£100,000
6	Maximum Loan	£1,000,000
7	Arrangement Fee Spread	0.5–1.0%
8	Minimum Length of Lease Required	40 years
9	Usual Length of Loan	2–20 years
10	Investment	Yes
11	Refinance	Yes
12	Owner Occupier	Yes
13	Residential Investment	Yes
14	Development Finance	No
15	Residential Development	No
16	Hotels and Leisure	No
17	Healthcare	No
18	Bridging and/or Mezzanine Finance	No

Merrill Lynch International Ltd

25 Ropemaker Street
London
EC2Y 9LY

Tel 020 7995 2787
Fax 020 7867 2008
www.ml.com

Contact
Naji E Boutros

Criteria

1	Maximum Loan to Value	100%
2	Benchmark Lending Rate	Varies
3	Margin Spread	Varies
4	Fixed	No
5	Minimum Loan	£30,000,000
6	Maximum Loan	None
7	Arrangement Fee Spread	Varies
8	Minimum Length of Lease Required	N/A
9	Usual Length of Loan	Varies
10	Investment	Yes
11	Refinance	Yes
12	Owner Occupier	Not usually
13	Residential Investment	In exceptional circumstances
14	Development Finance	Yes, pre-let
15	Residential Development	Rarely
16	Hotels and Leisure	Yes
17	Healthcare	Yes
18	Bridging and/or Mezzanine Finance	Yes

Remarks

Lends only to large plcs and major property players. Will lend or arrange loans worldwide. Prefers to lend on large portfolios, including leisure. Large refinancing of excisting holdings also possible.

Monmouthshire Building Society

Monmouthshire House
John Frost Square
Newport
South Wales
NP20 1PX

Tel 01633 844444
Fax 01633 844445
www.monbs.com

Contact
Collins Strong
Mortgage Lending Manager

Criteria

1	Maximum Loan to Value	80%
2	Benchmark Lending Rate	Base
3	Margin Spread	1.0–2.0%
4	Fixed	Requirement for hedging
5	Minimum Loan	£50,000
6	Maximum Loan	£1,000,000
7	Arrangement Fee Spread	1.0–3.0%
8	Minimum Length of Lease Required	70 years
9	Usual Length of Loan	3–30 years
10	Investment	Yes
11	Refinance	Yes
12	Owner Occupier	Yes
13	Residential Investment	Yes
14	Development Finance	No
15	Residential Development	No
16	Hotels and Leisure	No
17	Healthcare	No
18	Bridging and/or Mezzanine Finance	No

JPMorgan & Co

60 Wall Street
New York
New York 10260-0060
USA

Tel 001 212 483 2323

Contact
Jacques Thunnissen
Managing Director

Criteria

1	Maximum Loan to Value	100%
2	Benchmark Lending Rate	Cost of funds
3	Margin Spread	Varies
4	Fixed	Yes
5	Minimum Loan	£25,000,000
6	Maximum Loan	None
7	Arrangement Fee Spread	Varies
8	Minimum Length of Lease Required	N/A
9	Usual Length of Loan	25 years
10	Investment	Yes
11	Refinance	Yes
12	Owner Occupier	Yes
13	Residential Investment	Yes
14	Development Finance	Yes
15	Residential Development	Yes
16	Hotels and Leisure	Yes
17	Healthcare	Yes
18	Bridging and/or Mezzanine Finance	Yes

Remarks

Lends to major property companies and established investors and developers.

Morgan Guaranty Trust Co (JPMorgan)

60 Victoria Emankment
London
EC4Y 0JP

Tel 020 7600 2300
Fax 020 7325 5906

Contact
Linda Fussell

Mark Danzel

Criteria

1	Maximum Loan to Value	Variable
2	Benchmark Lending Rate	Base
3	Margin Spread	Negotiable
4	Fixed	No
5	Minimum Loan	£20,000,000
6	Maximum Loan	None
7	Arrangement Fee Spread	Variable
8	Minimum Length of Lease Required	None
9	Usual Length of Loan	1–10 years
10	Investment	Sometimes
11	Refinance	Sometimes
12	Owner Occupier	No
13	Residential Investment	No
14	Development Finance	No
15	Residential Development	No
16	Hotels and Leisure	Yes
17	Healthcare	Yes
18	Bridging and/or Mezzanine Finance	Yes

Remarks

Primarily funding for hotel and leisure industry.

Morgan Stanley Dean Witter

25 Cabot Street
Canary Wharf
London
E14 1QA

Tel 020 7425 5654
Fax 020 7425 3834

Contact
John A Carrafiell
Managing Director

Criteria

1	Maximum Loan to Value	95%
2	Benchmark Lending Rate	Fixed rate
3	Margin Spread	1.0–1.9%
4	Fixed	Yes — for term of loan
5	Minimum Loan	£10,000,000
6	Maximum Loan	None
7	Arrangement Fee Spread	0.5%
8	Minimum Length of Lease Required	N/A
9	Usual Length of Loan	Max 7 years
10	Investment	Yes
11	Refinance	Yes
12	Owner Occupier	Yes
13	Residential Investment	Yes
14	Development Finance	No
15	Residential Development	No
16	Hotels and Leisure	Yes
17	Healthcare	No
18	Bridging and/or Mezzanine Finance	No

Remarks

36 hours' credit approval from aggreement of terms.

Mortgage Trust Ltd

St Catherine's Court
Herbert Road
Solihull
West Midlands
B91 3QE

Tel 0870 4112323
Fax 01372 737777
www.mortgagetrust.co.uk

Contact
Austin Jelfs

Criteria

1	Maximum Loan to Value	90%
2	Benchmark Lending Rate	Libor
3	Margin Spread	1.0–2.0%
4	Fixed	Yes
5	Minimum Loan	£250,000
6	Maximum Loan	None
7	Arrangement Fee Spread	1.0–3.00%
8	Minimum Length of Lease Required	Varies
9	Usual Length of Loan	1–15 years
10	Investment	Yes
11	Refinance	Yes
12	Owner Occupier	Yes
13	Residential Investment	Yes
14	Development Finance	Yes
15	Residential Development	Yes
16	Hotels and Leisure	Yes
17	Healthcare	Yes
18	Bridging and/or Mezzanine Finance	No

Münchener Hypothekenbank eG

Karl-Scharnagl-Ring 10
80503 München
Germany

www.muenchener-hyp.de

Contact
Jan Polland
Head of Commercial Direct Busines
0895387 628
jan.polland@munchenerhyp.de

Criteria

1	Maximum Loan to Value	90%
2	Benchmark Lending Rate	Libor
3	Margin Spread	Varies
4	Fixed	Yes
5	Minimum Loan	£1,000,000
6	Maximum Loan	None
7	Arrangement Fee Spread	0.5–1.0%
8	Minimum Length of Lease Required	Varies
9	Usual Length of Loan	1–20 years
10	Investment	Yes
11	Refinance	Yes
12	Owner Occupier	No
13	Residential Investment	No
14	Development Finance	Yes
15	Residential Development	Yes
16	Hotels and Leisure	Yes
17	Healthcare	Yes
18	Bridging and/or Mezzanine Finance	Yes

National Counties Building Society

30 Church Street
Epsom
Surrey
KT17 4NL

Tel 01372 742211
Fax 01372 745607
www.ncbs.co.uk

Contact
Steve Iles
Head of Commercial Lending
01372 845380
commercial@ncbs.co.uk

Stewart Glower
Manager Special Schemes and Agreements
01372 845337
info@ncbs.co.uk

Criteria

1	Maximum Loan to Value	85%
2	Benchmark Lending Rate	Bank of England Bank rate
3	Margin Spread	0.5–3.0%
4	Fixed	Yes
5	Minimum Loan	£30,000
6	Maximum Loan	£10,000,000
7	Arrangement Fee Spread	1.0% negotiable
8	Minimum Length of Lease Required	50 years (30 central London at end of term)
9	Usual Length of Loan	Up to 25 years
10	Investment	Yes
11	Refinance	Yes
12	Owner Occupier	Yes
13	Residential Investment	Yes
14	Development Finance	No
15	Residential Development	No
16	Hotels and Leisure	Yes
17	Healthcare	Yes
18	Bridging and/or Mezzanine Finance	No

Nationwide Building Society

Kings Park Road
Moulton Park
Northampton
NN3 6NW

Tel 01604 854113
www.nationwide.co.uk

Contact
Nina Ghellenor
01604 854528

Julie Ferfield
General Manager — Birmingham
0121 625 0130

Criteria

1	Maximum Loan to Value	80%
2	Benchmark Lending Rate	Libor
3	Margin Spread	Libor 2.5–3.5%
4	Fixed	Yes
5	Minimum Loan	£50,000
6	Maximum Loan	None
7	Arrangement Fee Spread	1.0%
8	Minimum Length of Lease Required	30 years remaining at end of mortgage
9	Usual Length of Loan	5–25 years
10	Investment	Yes
11	Refinance	Yes
12	Owner Occupier	Yes
13	Residential Investment	Yes
14	Development Finance	No
15	Residential Development	No
16	Hotels and Leisure	No
17	Healthcare	No
18	Bridging and/or Mezzanine Finance	No

Remarks

With 15 key regional offices, we are sure to have the local knowledge you need.
Just scroll over the map to find the details of your nearest Commercial Finance team at www.nationwide.co.uk/commercial.

National Westminster Bank Plc

Structured Finance
Property Finance Group
5th Floor
Princess Street
London
EC2R 8PB

Tel 020 7390 1213
Fax 020 7390 1767
www.natwest.com

Contact
Stephen Eighteen

Criteria

1	Maximum Loan to Value	90%
2	Benchmark Lending Rate	Libor/cost of funds/base
3	Margin Spread	Varies
4	Fixed	5–25 years
5	Minimum Loan	£10,000,000
6	Maximum Loan	None
7	Arrangement Fee Spread	Negotiable
8	Minimum Length of Lease Required	None
9	Usual Length of Loan	1–25 years
10	Investment	Yes
11	Refinance	Yes
12	Owner Occupier	Yes
13	Residential Investment	Yes
14	Development Finance	Yes
15	Residential Development	Yes
16	Hotels and Leisure	Yes
17	Healthcare	Yes
18	Bridging and/or Mezzanine Finance	Yes

Newbury Building Society

17 Bartholomew Street
Newbury
Berkshire
RG14 5LY

Tel 01635 555700
Fax 01635 555799
www.newbury.co.uk

Contact
Phillippa Cardno
Assistant General Manager
of Customer Services

Criteria

1	Maximum Loan to Value	85%
2	Benchmark Lending Rate	Lender base
3	Margin Spread	0.5–2.0%
4	Fixed	No
5	Minimum Loan	£75,000
6	Maximum Loan	£2,000,000
7	Arrangement Fee Spread	0.25–1.0%
8	Minimum Length of Lease Required	30 years plus length of loan
9	Usual Length of Loan	5–20 years
10	Investment	Yes
11	Refinance	Yes
12	Owner Occupier	Yes
13	Residential Investment	Yes
14	Development Finance	No
15	Residential Development	No
16	Hotels and Leisure	Yes
17	Healthcare	Yes
18	Bridging and/or Mezzanine Finance	No

Remarks

All types of quality property funded within a 60-mile radius of Newbury.

Newcastle Building Society

Portland House
New Bridge Street
Newcastle upon Tyne
NE1 8AL

Tel 0191 244 2295/1821/2256 or 0191 255 8357
Fax 0191 244 2271
www.newcastle.co.uk

Contact

Colin Barnes
Head of Property Finance
colin.barnes@newcastle.co.uk

Mike Patterson
Senior Manager Commercial Lendin

Sally Peverley
Senior Manager Commercial Lendin

Ken Kerr
Senior Manager Housing Finance

Criteria

1	Maximum Loan to Value	85%
2	Benchmark Lending Rate	3 month Libor
3	Margin Spread	0.80–1.50%
4	Fixed	Yes
5	Minimum Loan	£750,000
6	Maximum Loan	£25,000,000
7	Arrangement Fee Spread	0.35–1.0%
8	Minimum Length of Lease Required	Negotiable
9	Usual Length of Loan	5–20 years
10	Investment	Yes (core business)
11	Refinance	Yes
12	Owner Occupier	No
13	Residential Investment	Yes (core business)
14	Development Finance	No
15	Residential Development	No
16	Hotels and Leisure	Investment only
17	Healthcare	Investment only
18	Bridging and/or Mezzanine Finance	No

Remarks

An investment property lender with a strong emhpasis on quality property and borrower credentials to provide flexible terms.

Nomura Bank International

Nomura House
1 St Martin's-le-Grand
London
EC1A 4NP

Tel 020 7521 2000
Fax 020 7626 0851
www.nomura.com

Contact
J G S Macdonald

Charles Bowen

Guy Hands

Douglas Grant

Criteria

1	Maximum Loan to Value	100%
2	Benchmark Lending Rate	3 month Libor
3	Margin Spread	0.9–1.75%
4	Fixed	Yes
5	Minimum Loan	£10,000,000
6	Maximum Loan	None
7	Arrangement Fee Spread	0.25–1.0%
8	Minimum Length of Lease Required	Varies
9	Usual Length of Loan	Up to 5 years
10	Investment	Yes
11	Refinance	Yes
12	Owner Occupier	Yes
13	Residential Investment	No
14	Development Finance	No
15	Residential Development	No
16	Hotels and Leisure	No
17	Healthcare	No
18	Bridging and/or Mezzanine Finance	Yes

Remarks

Lends almost excluslively to major plcs and property companies.

Nordea

8th Floor
City Place House
55 Basinghall Street
London
EC2V 5NB

Tel 020 7726 9000
Fax 020 7726 9009

Contact
William Goldstein

Steve Ambridge

Criteria

1	Maximum Loan to Value	80%
2	Benchmark Lending Rate	Bank base/3 month Libor
3	Margin Spread	1.0–2.0%
4	Fixed	Yes
5	Minimum Loan	£500,000
6	Maximum Loan	£20,000,000
7	Arrangement Fee Spread	0.25–1.0%
8	Minimum Length of Lease Required	Loan period plus 5 years
9	Usual Length of Loan	Up to 10 years
10	Investment	Yes
11	Refinance	Yes
12	Owner Occupier	Only for clients of first-class financial standing
13	Residential Investment	As part of a mixed portfolio
14	Development Finance	Yes, with pre-let
15	Residential Development	No
16	Hotels and Leisure	Yes, subject to operator's lease
17	Healthcare	No
18	Bridging and/or Mezzanine Finance	No

Remarks

Formely Merita Nordbanken.

Northern Bank Ltd

Donegal Square West
Belfast
BT1 6JS

Tel 01232 245277
Fax 01232 893337
www.northernbank.co.uk

Contact
Winston Fletcher

Criteria

1	Maximum Loan to Value	85%
2	Benchmark Lending Rate	Base
3	Margin Spread	1.0–2.0%
4	Fixed	No
5	Minimum Loan	£15,000
6	Maximum Loan	None
7	Arrangement Fee Spread	0.25–1.0%
8	Minimum Length of Lease Required	None
9	Usual Length of Loan	1 month–30 years
10	Investment	Yes
11	Refinance	Yes
12	Owner Occupier	Yes
13	Residential Investment	Yes
14	Development Finance	Yes
15	Residential Development	Yes
16	Hotels and Leisure	Yes
17	Healthcare	Yes
18	Bridging and/or Mezzanine Finance	No

Northern Rock Plc

Northern Rock Commercial Finance
Doxford International Business Park
Sunderland
SR3 3XG

Tel 0191 522 3777
Fax 0191 522 3750/3751
www.northernrock.co.uk

Contact

Christina Foley
Newcastle
0191 221 0383

Nigel Sims
Manchester
0161 843 6484

Grayam Powel
Birmingham
0121 643 5153

Criteria

1	Maximum Loan to Value	85%
2	Benchmark Lending Rate	Building society base rate/ 3 month Libor/ commercial tracker
3	Margin Spread	1.5–2.5%
4	Fixed	Yes, subject to availability
5	Minimum Loan	£30,000 £100,000 offices, retail £250,000 nursing homes
6	Maximum Loan	£10,000,000
7	Arrangement Fee Spread	0.5–1.0%
8	Minimum Length of Lease Required	Mortgage term plus 30 years
9	Usual Length of Loan	5–25 years
10	Investment	Yes
11	Refinance	Yes
12	Owner Occupier	Yes
13	Residential Investment	Yes
14	Development Finance	Pre-let/owner-occupied commercial development considered
15	Residential Development	No
16	Hotels and Leisure	Yes
17	Healthcare	Yes
18	Bridging and/or Mezzanine Finance	No

Remarks

The government has announced the nationalisation of the Northern Rock and it is not clear what, if any, changes there will be to its criteria. The criteria above were as given before the current events.

Norwich & Peterborough Building Society

Peterborough Business Park
Lynch Wood
Peterborough
PE2 6WZ

Tel 01733 372372
Fax 01733 372373
www.npbs.co.uk

Contact
Stuart Benge
01733 372242
stuart.benge@npbs.co.uk

Criteria

1	Maximum Loan to Value	85%
2	Benchmark Lending Rate	Libor/mortgage base
3	Margin Spread	0.5–3.0% (Libor) 0–2.0% (base)
4	Fixed	Available — please enquire
5	Minimum Loan	£100,000
6	Maximum Loan	£10,000,000
7	Arrangement Fee Spread	0.5–1.0%
8	Minimum Length of Lease Required	50 years plus at the end of loan on long leasehold
9	Usual Length of Loan	5–25 years
10	Investment	Yes
11	Refinance	Yes
12	Owner Occupier	Yes
13	Residential Investment	No
14	Development Finance	No
15	Residential Development	No
16	Hotels and Leisure	Yes
17	Healthcare	Yes
18	Bridging and/or Mezzanine Finance	No

Norwich Union

Surrey Street
Norwich
NR1 3NS

Tel 01603 622200
Fax 01603 683659
www.norwichunion.com

Contact
Allen Collett
Commercial Mortgage Manager
01603 684 907

Criteria

1	Maximum Loan to Value	90%
2	Benchmark Lending Rate	Gilts
3	Margin Spread	1.6–2.0%
4	Fixed	Yes
5	Minimum Loan	£1,000,000
6	Maximum Loan	None
7	Arrangement Fee Spread	0.15% of loan amount
8	Minimum Length of Lease Required	At least 35 years unexpired at the end of the mortgage term
9	Usual Length of Loan	5–25 years
10	Investment	Yes
11	Refinance	Yes
12	Owner Occupier	No
13	Residential Investment	No
14	Development Finance	No
15	Residential Development	No
16	Hotels and Leisure	No
17	Healthcare	No
18	Bridging and/or Mezzanine Finance	No

Nottingham Building Society

Nottingham House
5-13 Upper Parliament Street
Nottingham
NG1 2BX

Tel 0115 948 1444
www.thenottingham.com

Contact
Gary Burrows
Commercial Lending Manager
0115 956 4653
gary.burrows@thenottingham.com

Criteria

1	Maximum Loan to Value	80%
2	Benchmark Lending Rate	Commercial mortgage rate
3	Margin Spread	1.0–3.0%
4	Fixed	N/A
5	Minimum Loan	£50,000
6	Maximum Loan	£1,000,000
7	Arrangement Fee Spread	Minimum 1.0%
8	Minimum Length of Lease Required	Term of loan + 30 years
9	Usual Length of Loan	3–30 years
10	Investment	Yes
11	Refinance	Yes
12	Owner Occupier	Yes
13	Residential Investment	Yes
14	Development Finance	No
15	Residential Development	No
16	Hotels and Leisure	Yes
17	Healthcare	Yes
18	Bridging and/or Mezzanine Finance	No

Paragon Mortgages Ltd

St Catherine's Court
Herbert Road
Solihull
West Midlands
B91 3QE

Tel 0800 440099
Fax 0121 712 2532
www.paragon-mortgages.co.uk

Contact
John Heron
Managing Director
0121 712 256 / Ext: 2561

Criteria

1	Maximum Loan to Value	90%
2	Benchmark Lending Rate	Libor
3	Margin Spread	1.0–2.0%
4	Fixed	Yes
5	Minimum Loan	£500,000
6	Maximum Loan	£10,000,000
7	Arrangement Fee Spread	1.0–3.00%
8	Minimum Length of Lease Required	40 years
9	Usual Length of Loan	1–10 years
10	Investment	Yes
11	Refinance	Yes
12	Owner Occupier	Yes
13	Residential Investment	Yes
14	Development Finance	Yes
15	Residential Development	Yes
16	Hotels and Leisure	Yes
17	Healthcare	Yes
18	Bridging and/or Mezzanine Finance	Yes

Principality Building Society

Principality Buildings
Queen Street
Cardiff
CF10 1UA

Tel 0845 045 0006
Fax 029 2077 3412
www.principality.co.uk

Contact
Peter Hughes
Head of Commercial Lending

Jon Ursell
Commercial Lending Manager

Ian Bailey
Commercial Lending Manager

Gareth Redding
Commercial Lending Manager

Criteria

1	Maximum Loan to Value	85%
2	Benchmark Lending Rate	Standard variable rate
3	Margin Spread	Up to 2%
4	Fixed	Residential only
5	Minimum Loan	£15,000
6	Maximum Loan	£1,000,000
7	Arrangement Fee Spread	0.75–1.25%
8	Minimum Length of Lease Required	30 years
9	Usual Length of Loan	5–25 years
10	Investment	Yes
11	Refinance	Yes
12	Owner Occupier	Yes
13	Residential Investment	Yes
14	Development Finance	Yes
15	Residential Development	Yes
16	Hotels and Leisure	No
17	Healthcare	No
18	Bridging and/or Mezzanine Finance	No

Remarks

Lending in Wales only.

Progressive Building Society

33/37 Wellington Place
Belfast
BT1 6HH

Tel 028 90244926
Fax 028 90330431
www.theprogressive.com

Contact
Peter Lyttle
Mortgage Account Manager

Criteria

1	Maximum Loan to Value	80%
2	Benchmark Lending Rate	Base
3	Margin Spread	1.0–3.0%
4	Fixed	Yes
5	Minimum Loan	£75,000
6	Maximum Loan	£1,000,000
7	Arrangement Fee Spread	1.0–2.00%
8	Minimum Length of Lease Required	35 years
9	Usual Length of Loan	5–25 years
10	Investment	Yes
11	Refinance	Yes
12	Owner Occupier	Yes
13	Residential Investment	Yes
14	Development Finance	No
15	Residential Development	No
16	Hotels and Leisure	No
17	Healthcare	No
18	Bridging and/or Mezzanine Finance	No

Prudential (Portfolio Managers Ltd)

142 Holborn Bars
London
EC1N 2NH

Tel 020 7548 3089/3936
Fax 020 7548 3989

Contact
Steven Chamberlain

Dennis Collins

Martin Walsh

Criteria

1	Maximum Loan to Value	85%
2	Benchmark Lending Rate	3 month Libor
3	Margin Spread	1.0–2.0%
4	Fixed	Yes, up to 30 years
5	Minimum Loan	£1,000,000
6	Maximum Loan	None
7	Arrangement Fee Spread	0.25–0.5%
8	Minimum Length of Lease Required	70 years
9	Usual Length of Loan	15–30 years
10	Investment	Yes
11	Refinance	Yes
12	Owner Occupier	Yes
13	Residential Investment	No
14	Development Finance	No
15	Residential Development	No
16	Hotels and Leisure	Yes
17	Healthcare	Yes
18	Bridging and/or Mezzanine Finance	No

Remarks

Specialises in lending long-term on fixed rate on all types of UK quality commercial property.

Regentsmead Ltd

Russel House
140 High Street
Edgware
Middlesex
HA8 7LW

Tel 020 8952 1414
Fax 020 8952 2424
www.regentsmead.com
info@regentsmead.com

Contact
James Bloom
Chief Executive
jbloom@regentsmead.com

Nicholas Warren
Senior Executive
nwarren@regentsmead.com

Criteria

1	Maximum Loan to Value	70–80% costs
2	Benchmark Lending Rate	1.25% per month
3	Margin Spread	1.0–1.5% per month
4	Fixed	Yes
5	Minimum Loan	£100,000
6	Maximum Loan	£6,000,000
7	Arrangement Fee Spread	1.0%
8	Minimum Length of Lease Required	90 years
9	Usual Length of Loan	Up to 12 months
10	Investment	No
11	Refinance	Yes
12	Owner Occupier	No, as not regulated by FSA
13	Residential Investment	No
14	Development Finance	Yes
15	Residential Development	Yes
16	Hotels and Leisure	No
17	Healthcare	No
18	Bridging and/or Mezzanine Finance	Bridging finance only

Remarks

Very keen on quality development finance but will look at all short-term funding required. Aims to issue offer in principle within 24 hours of receiving suitable proposal.

N M Rothschild & Sons

New Court
St Swithin's Lane
London
EC4P 4DU

Tel 020 7280 5000
Fax 020 7280 1918
www.rothschild.co.uk

Contact

Charles Keay
Head of Real Estate Finance

Adam Greenbury
Head of Real Estate Finance — London Team

Criteria

1	Maximum Loan to Value	90%
2	Benchmark Lending Rate	Libor
3	Margin Spread	1.0–2.0%
4	Fixed	Yes
5	Minimum Loan	£2,000,000
6	Maximum Loan	None
7	Arrangement Fee Spread	0.25–1.05%
8	Minimum Length of Lease Required	No
9	Usual Length of Loan	5 years
10	Investment	Yes
11	Refinance	Yes
12	Owner Occupier	No
13	Residential Investment	No
14	Development Finance	Yes
15	Residential Development	No
16	Hotels and Leisure	No
17	Healthcare	No
18	Bridging and/or Mezzanine Finance	Yes

The Royal Bank of Scotland

Structured Property Finance
Level 5
135 Bishopsgate
London
EC2M 3UR

Contact

Paul Carpenter
Head of Structured Property Finance
020 7085 5000

Stephen Eighteen
Structured Property Group
020 7085 3930

Criteria

1	Maximum Loan to Value	Varies
2	Benchmark Lending Rate	Libor/cost of funds/base
3	Margin Spread	Varies
4	Fixed	2–25 years
5	Minimum Loan	£10,000,000
6	Maximum Loan	None
7	Arrangement Fee Spread	Negotiable
8	Minimum Length of Lease Required	None
9	Usual Length of Loan	1–25 years
10	Investment	Yes
11	Refinance	Yes
12	Owner Occupier	Yes
13	Residential Investment	Yes
14	Development Finance	Yes — pre-let only
15	Residential Development	Yes
16	Hotels and Leisure	Yes
17	Healthcare	Yes
18	Bridging and/or Mezzanine Finance	No

Remarks

Active in providing a range of funding solutions, including joint venture equity, mezzanine and senior debt.

Saffron Walden Herts & Essex Building Society

Saffron House
1A Market Street
Saffron Walden
Essex
CB10 1HX

Tel 01799 522211
Fax 01799 513622
www.saffronbs.co.uk

Contact
Tom Thomas
Head of Mortgage Underwriting

Criteria

1	Maximum Loan to Value	85%
2	Benchmark Lending Rate	Base rate
3	Margin Spread	1.0–2.5%
4	Fixed	No
5	Minimum Loan	£25,000
6	Maximum Loan	£500,000
7	Arrangement Fee Spread	0–1.5%
8	Minimum Length of Lease Required	50 years following term expiry
9	Usual Length of Loan	25 years
10	Investment	Yes
11	Refinance	Yes
12	Owner Occupier	Yes
13	Residential Investment	Yes
14	Development Finance	Yes
15	Residential Development	Yes
16	Hotels and Leisure	Yes
17	Healthcare	No
18	Bridging and/or Mezzanine Finance	No

Remarks

Prefers lending on residential investment and developments.

Sainsbury's Bank

City House
City Road
Chester
CH88 3YE

Tel 0800 096 0541
www.sainsburysbank.co.uk

Contact
Peter Wood
Director of Savings and Mortgages

Criteria

1	Maximum Loan to Value	75%
2	Benchmark Lending Rate	Libor
3	Margin Spread	1.0%
4	Fixed	Yes
5	Minimum Loan	£25,001
6	Maximum Loan	£1,000,000
7	Arrangement Fee Spread	1.0%
8	Minimum Length of Lease Required	N/A
9	Usual Length of Loan	5–25 years
10	Investment	No
11	Refinance	No
12	Owner Occupier	No
13	Residential Investment	Yes
14	Development Finance	No
15	Residential Development	Yes
16	Hotels and Leisure	No
17	Healthcare	No
18	Bridging and/or Mezzanine Finance	No

Scottish Agricultural Securities Corporation Plc

29 St Vincent Place
Glasgow
G1 2EU

Tel 0141 223 2598

Contact
John Mirtle

Criteria

1	Maximum Loan to Value	66%
2	Benchmark Lending Rate	Cost of fixed rate funds
3	Margin Spread	Varies
4	Fixed	Yes
5	Minimum Loan	£15,001
6	Maximum Loan	£2,000,000
7	Arrangement Fee Spread	None
8	Minimum Length of Lease Required	Freehold only
9	Usual Length of Loan	5–25 years
10	Investment	No
11	Refinance	Yes
12	Owner Occupier	Yes
13	Residential Investment	No
14	Development Finance	No
15	Residential Development	No
16	Hotels and Leisure	No
17	Healthcare	No
18	Bridging and/or Mezzanine Finance	No

Remarks

Loans on agricultural properties only. Lending exclusively to existing clients only.

Singer & Friedlander Ltd

One Hanover Street
London
W1S 1AX

Tel 020 3205 5000
Fax 020 3205 5001
www.kaupthingsingers.co.uk

Contact
Andrew Gent
Director, Corporate and
Structured Finance
020 3205 5105

Dan Walker
Assistant Director
020 3205 5117

Matthew Male
Manager — Manchester
0161 819 4493

Steve Lewis
Director, Corporate & Structured
Finance — Birmingham
0121 200 3721

Criteria

1	Maximum Loan to Value	90%
2	Benchmark Lending Rate	3 month Libor
3	Margin Spread	1.8–2.0%
4	Fixed	Sometimes
5	Minimum Loan	£1,000,000
6	Maximum Loan	£20,000,000
7	Arrangement Fee Spread	0.5–1.0%
8	Minimum Length of Lease Required	70 years
9	Usual Length of Loan	1–2 years
10	Investment	Yes
11	Refinance	Yes
12	Owner Occupier	Yes
13	Residential Investment	Yes
14	Development Finance	Yes
15	Residential Development	Yes
16	Hotels and Leisure	Yes
17	Healthcare	Yes
18	Bridging and/or Mezzanine Finance	Yes

Remarks

Prefers to lend in the development sector — will also fund commercial refurbishment. Will look at residential development funding or refurbishment of office buildings, residential in large city centres only. No speculative commercial development finance. 100% development packages considered. Preferred size for these: £1m–£5m. Investment roll-up possible.

Skipton Building Society

The Bailey
Skipton
North Yorkshire
BD23 1DN

Tel 01756 705027
Fax 01756 705729
www.skipton-commercial.co.uk
commlend@skipton.co.uk

Contact
Commercial Mortgage Manager
at nearest branch

Criteria

1	Maximum Loan to Value	75% 80% for blue-chip investment
2	Benchmark Lending Rate	Base
3	Margin Spread	Bank base + 0.85–2.25%
4	Fixed	6.99% — 5 years
5	Minimum Loan	£25,000
6	Maximum Loan	£10,000,000
7	Arrangement Fee Spread	£400–4,000 up to £1,000,000 then negotiable
8	Minimum Length of Lease Required	Negotiable
9	Usual Length of Loan	5–25 years
10	Investment	Yes
11	Refinance	Yes
12	Owner Occupier	Yes
13	Residential Investment	Yes
14	Development Finance	No
15	Residential Development	No
16	Hotels and Leisure	Yes
17	Healthcare	Yes
18	Bridging and/or Mezzanine Finance	No

Société Générale

Corporate and Investment Banking
Exchange House
Primrose Street
Broadgate
London
EC2A 2HT

Tel 020 7676 6000
Fax 020 7702 4424
www.sgcib.com

Contact
Patrick Clonrozier
patrick.clonrozier@sgcib.com

Criteria

1	Maximum Loan to Value	75%
2	Benchmark Lending Rate	3 month Libor
3	Margin Spread	0.85–1.5%
4	Fixed	Yes
5	Minimum Loan	£5,000,000
6	Maximum Loan	None
7	Arrangement Fee Spread	0.25–0.5%
8	Minimum Length of Lease Required	15 yeas
9	Usual Length of Loan	5–7 years
10	Investment	Yes
11	Refinance	Yes
12	Owner Occupier	No
13	Residential Investment	No
14	Development Finance	Yes
15	Residential Development	No
16	Hotels and Leisure	Yes
17	Healthcare	No
18	Bridging and/or Mezzanine Finance	No

Remarks

All types of commercial property, including pre-let development situations. Particulary keen on portfolios as well as corporate deals linked to equity issues.

Stroud & Swindon

Rowcroft
Stroud
Gloucestershire
GL5 3BG

Tel 01453 757011
www.stroudandswindon.co.uk

Contact
Andy King
Senior Manager of Lending

Criteria

1	Maximum Loan to Value	75%
2	Benchmark Lending Rate	Base
3	Margin Spread	1–2.5%
4	Fixed	Yes
5	Minimum Loan	£125,000
6	Maximum Loan	None
7	Arrangement Fee Spread	1%
8	Minimum Length of Lease Required	45 years
9	Usual Length of Loan	5–25 years
10	Investment	Yes
11	Refinance	Yes
12	Owner Occupier	No
13	Residential Investment	No
14	Development Finance	No
15	Residential Development	No
16	Hotels and Leisure	No
17	Healthcare	No
18	Bridging and/or Mezzanine Finance	No

Sumitomo Bank

99 Queen Victoria Street
London
EC4V 4EH

Tel 020 7786 1000
Fax 020 7236 0049
www.smbc.co.jp

Contact
Alan Springett

Kevin Harmer

Criteria

1	Maximum Loan to Value	75%
2	Benchmark Lending Rate	Cost of funds
3	Margin Spread	Varies
4	Fixed	Yes
5	Minimum Loan	£2,000,000
6	Maximum Loan	None
7	Arrangement Fee Spread	Varies
8	Minimum Length of Lease Required	None
9	Usual Length of Loan	Varies
10	Investment	Yes
11	Refinance	Yes
12	Owner Occupier	No
13	Residential Investment	No
14	Development Finance	Yes, pre-let
15	Residential Development	No
16	Hotels and Leisure	No
17	Healthcare	No
18	Bridging and/or Mezzanine Finance	No

Svenska Handelsbanken

Trinity Tower
9 Thomas More Street
London
E1W 1WY

Tel 020 7578 8000
Fax 020 7578 8004
www.handelsbanken.co.uk

Contact
Lars Backlund
Head of Asset Finance
020 7578 8101

Criteria

1	Maximum Loan to Value	75%
2	Benchmark Lending Rate	Bank base/3 month Libor
3	Margin Spread	1.5–2.5%
4	Fixed	Yes, 5 years
5	Minimum Loan	£500,000
6	Maximum Loan	£10,000,000
7	Arrangement Fee Spread	0.25–0.75%
8	Minimum Length of Lease Required	100 years
9	Usual Length of Loan	5–7 years
10	Investment	Yes
11	Refinance	Yes
12	Owner Occupier	Yes
13	Residential Investment	No
14	Development Finance	No
15	Residential Development	No
16	Hotels and Leisure	No
17	Healthcare	No
18	Bridging and/or Mezzanine Finance	No

Remarks

Prefers lending to clients with a Scandinavian connection.

Swansea Building Society

11 & 12 Cradock Street
Swansea
SA1 3EW

Tel 01792 483701
Fax 01792 483718
www.swansea-bs.co.uk

Contact
Allen Williamson
Chief Executive
mortgage@swansea-bs.co.uk

Criteria

1	Maximum Loan to Value	80%
2	Benchmark Lending Rate	Base
3	Margin Spread	1.0–2.0%
4	Fixed	Yes
5	Minimum Loan	£25,000
6	Maximum Loan	£5,000,000
7	Arrangement Fee Spread	0.5–1.0%
8	Minimum Length of Lease Required	45 years
9	Usual Length of Loan	Up to 30 years
10	Investment	Yes
11	Refinance	Yes
12	Owner Occupier	Yes
13	Residential Investment	Yes
14	Development Finance	No
15	Residential Development	No
16	Hotels and Leisure	Yes
17	Healthcare	No
18	Bridging and/or Mezzanine Finance	No

The Beverly Building Society

57 Market Place
Beverley
East Yorkshire
HU17 8AA

Tel 01482 881510
Fax 01482 872680
www.beverleybs.co.uk

Contact
Mortgage Department
01482 881510

Criteria

1	Maximum Loan to Value	75%
2	Benchmark Lending Rate	Base
3	Margin Spread	1.0–2.0%
4	Fixed	Ye
5	Minimum Loan	£50,000
6	Maximum Loan	£2,500,000
7	Arrangement Fee Spread	1.0–2.0%
8	Minimum Length of Lease Required	50 years
9	Usual Length of Loan	5–25 years
10	Investment	Yes
11	Refinance	Yes
12	Owner Occupier	Yes
13	Residential Investment	Yes
14	Development Finance	No
15	Residential Development	No
16	Hotels and Leisure	No
17	Healthcare	No
18	Bridging and/or Mezzanine Finance	No

The Chorley Building Society

Key House
Foxhole Road
Chorley
PR7 1NZ

Tel 01257 279373
Fax 01257 241371
www.chorleybs.co.uk

Contact

Mortgage Team
01257 419110
mortgages@chorleybs.co.uk

Criteria

1	Maximum Loan to Value	75%
2	Benchmark Lending Rate	Base
3	Margin Spread	1.0–2.0%
4	Fixed	Yes
5	Minimum Loan	£100,000
6	Maximum Loan	£1,000,000
7	Arrangement Fee Spread	1.0–3.00%
8	Minimum Length of Lease Required	40 years
9	Usual Length of Loan	3–30 years
10	Investment	Yes
11	Refinance	Yes
12	Owner Occupier	Yes
13	Residential Investment	Yes
14	Development Finance	No
15	Residential Development	No
16	Hotels and Leisure	No
17	Healthcare	No
18	Bridging and/or Mezzanine Finance	No

The Hanley Economic Building Society

Granville House
Festival Park
Hanley
Stoke-on-Trent
ST1 5TB

Tel 0800 542 8790
Fax 01782 873161
www.thehanley.co.uk

Contact
Diane Thompson
Head of Retail Opertions
01782 255000

Criteria

1	Maximum Loan to Value	80%
2	Benchmark Lending Rate	Base
3	Margin Spread	1.0–3.0%
4	Fixed	Yes
5	Minimum Loan	£75,000
6	Maximum Loan	£2,500,000
7	Arrangement Fee Spread	0.5–1.0%
8	Minimum Length of Lease Required	40 years
9	Usual Length of Loan	3–15 years
10	Investment	Yes
11	Refinance	Yes
12	Owner Occupier	Yes
13	Residential Investment	Yes
14	Development Finance	No
15	Residential Development	No
16	Hotels and Leisure	No
17	Healthcare	No
18	Bridging and/or Mezzanine Finance	No

The Mortgage Works

www.themortgageworks.co.uk

Contact
Paula Priest
Regional Sales Manager
(Midlands)
07889 829443
paula.priest@themortgageworks.co

Jayne Wilde
Regional Sales Manager
(North)
07887 830067
jayne.wilde@themortgageworks.co.

Zoe Griffin
Regional Sales Manager
(South West)
07725 680011
zoe.griffin@ucbhomeloans.co.uk

John Dwyer
Regional Sales Manager
(South East)
07725 680010
john.dwyer@ucbhomeloans.co.uk

Criteria

1	Maximum Loan to Value	85%
2	Benchmark Lending Rate	Libor
3	Margin Spread	1.0–3.0%
4	Fixed	Yes
5	Minimum Loan	£350,000
6	Maximum Loan	£5,000,000
7	Arrangement Fee Spread	1.0–3.0%
8	Minimum Length of Lease Required	Varies
9	Usual Length of Loan	1–10 years
10	Investment	Yes
11	Refinance	Yes
12	Owner Occupier	Yes
13	Residential Investment	Yes
14	Development Finance	Yes
15	Residential Development	Yes
16	Hotels and Leisure	Yes
17	Healthcare	Yes
18	Bridging and/or Mezzanine Finance	Yes

UBS

Global Asset Management
100 Liverpool Street
London
EC2M 2RH

Tel 020 7568 7726
020 7567 8000
Fax 020 7568 4800
020 7568 0406
www.ubs.com

Contact
Annette Alford

Criteria

1	Maximum Loan to Value	100%
2	Benchmark Lending Rate	Libor
3	Margin Spread	Varies
4	Fixed	Sometimes
5	Minimum Loan	£5,000,000
6	Maximum Loan	None
7	Arrangement Fee Spread	Varies
8	Minimum Length of Lease Required	N/A
9	Usual Length of Loan	5–15 years
10	Investment	Yes
11	Refinance	Yes
12	Owner Occupier	Yes
13	Residential Investment	No
14	Development Finance	No
15	Residential Development	No
16	Hotels and Leisure	No
17	Healthcare	No
18	Bridging and/or Mezzanine Finance	No

Ulster Bank Ltd

11-16 Donegall Square East
Belfast
BT1 5UB

Tel 0845 603 6402
www.ulsterbank.co.uk

Contact
Henry Elvin
Head of Corporate Banking
and Property Division
02890 276128
henry.elvin@ulsterbank.com

Criteria

1	Maximum Loan to Value	85%
2	Benchmark Lending Rate	Bank base
3	Margin Spread	2.0–4.0%
4	Fixed	Yes
5	Minimum Loan	£75,000
6	Maximum Loan	£5,000,000
7	Arrangement Fee Spread	0.75–1.0%
8	Minimum Length of Lease Required	Term of loan plus 40 years
9	Usual Length of Loan	1 month–25 years
10	Investment	Yes
11	Refinance	Yes
12	Owner Occupier	Yes
13	Residential Investment	Yes
14	Development Finance	Yes
15	Residential Development	Yes
16	Hotels and Leisure	Yes
17	Healthcare	Yes
18	Bridging and/or Mezzanine Finance	Yes

Remarks

Lends exclusively in Northern Ireland and the Irish Republic.

United Trust Bank Ltd

80 Haymarket
London
SW1Y 4TE

Tel 020 7190 5555
Fax 020 7190 5550
www.utbank.co.uk

Contact
Mike Campion
Business Development Manager

Noel Meredith
Business Development Manager

Criteria

1	Maximum Loan to Value	90%
2	Benchmark Lending Rate	Base rate
3	Margin Spread	2–5% (bridging finance 1–2% per month)
4	Fixed	No
5	Minimum Loan	£100,000
6	Maximum Loan	£2,500,000
7	Arrangement Fee Spread	1.0–2.0%
8	Minimum Length of Lease Required	Negotiable
9	Usual Length of Loan	1–5 years
10	Investment	Yes
11	Refinance	Yes, if component in a wider project
12	Owner Occupier	Yes, if component in a wider project
13	Residential Investment	Yes
14	Development Finance	Yes — commercial development
15	Residential Development	Yes
16	Hotels and Leisure	No
17	Healthcare	Yes
18	Bridging and/or Mezzanine Finance	Yes

Remarks

While the main focus is on development funding and bridging finance (including planning gains), the bank has developed a niche for shorter-term mezzanine/equity finanace.

Vernon Building Society

19 St Petersgate
Stockport
SK1 1HF

Tel 0161 429 6262
Fax 0161 429 6996
www.thevernon.co.uk

Contact
Mike Hanson
Chief Executive

Criteria

1	Maximum Loan to Value	85%
2	Benchmark Lending Rate	Standard variable rate
3	Margin Spread	1.0–2.0%
4	Fixed	No
5	Minimum Loan	£15,000
6	Maximum Loan	£1,000,000
7	Arrangement Fee Spread	1%
8	Minimum Length of Lease Required	N/A
9	Usual Length of Loan	10–20 years
10	Investment	Yes
11	Refinance	Yes
12	Owner Occupier	Yes
13	Residential Investment	Yes
14	Development Finance	Yes — pre-let only
15	Residential Development	No
16	Hotels and Leisure	No
17	Healthcare	No
18	Bridging and/or Mezzanine Finance	No

West Bromwich Building Society

374 High Street
West Bromwich
West Midlands
B70 8LR

Tel 0870 220 7744
Fax 0121 525 0572
www.westbrom.co.uk

Contact
M J Nixon
07970 130234

A Smith
07970 130213

A Faed
07852 906524

Criteria

1	Maximum Loan to Value	80%
2	Benchmark Lending Rate	Libor/Swap
3	Margin Spread	0.9–2.5%
4	Fixed	Yes
5	Minimum Loan	£250,000
6	Maximum Loan	£30,000,000
7	Arrangement Fee Spread	0.25–1.0%
8	Minimum Length of Lease Required	35 years remaining at end of loan
9	Usual Length of Loan	5–20 years
10	Investment	Yes
11	Refinance	Yes
12	Owner Occupier	No
13	Residential Investment	Yes
14	Development Finance	Sometimes
15	Residential Development	No
16	Hotels and Leisure	Yes
17	Healthcare	Yes
18	Bridging and/or Mezzanine Finance	No

Remarks

85% available for "blue chip" investment deals.

Westdeutsche Immobilien Bank

Woolgate Exchange
25 Basinghall Street
London
EC2V 5HA

Tel 020 7020 5400
Fax 020 7020 5401
www.westimmobank.com

Contact
Klaus Schreiner
Head of London Branch
020 7020 5401
klaus.schreiner@westimmobank.co

Jane Simpson
Head of Origination
020 7020 5402
jane.simpson@westimmobank.com

Shirley Sinclair
Director
020 7020 5403
shirley.sinclair@westimmobank.com

Criteria

1	Maximum Loan to Value	80%
2	Benchmark Lending Rate	Libor
3	Margin Spread	0.80–1.75%
4	Fixed	N/A
5	Minimum Loan	£5,000,000
6	Maximum Loan	£60,000,000
7	Arrangement Fee Spread	0.25–0.60%
8	Minimum Length of Lease Required	10 years
9	Usual Length of Loan	5–10 years
10	Investment	Yes
11	Refinance	Yes
12	Owner Occupier	No
13	Residential Investment	Yes
14	Development Finance	Yes
15	Residential Development	Yes
16	Hotels and Leisure	Yes
17	Healthcare	Yes
18	Bridging and/or Mezzanine Finance	No

Yorkshire Bank Plc

20 Merrion Way
Leeds
LS2 8NZ

Tel 0113 247 2000
Fax 0113 242 0733

Contact
Commercial Mortgage Manager

Criteria

1	Maximum Loan to Value	85%
2	Benchmark Lending Rate	Bank base
3	Margin Spread	Varies
4	Fixed	Yes
5	Minimum Loan	None
6	Maximum Loan	None
7	Arrangement Fee Spread	1.0–2.0%
8	Minimum Length of Lease Required	Varies
9	Usual Length of Loan	Up to 10 years
10	Investment	Yes
11	Refinance	Yes
12	Owner Occupier	Yes
13	Residential Investment	Yes
14	Development Finance	Yes
15	Residential Development	Yes
16	Hotels and Leisure	Yes
17	Healthcare	Yes
18	Bridging and/or Mezzanine Finance	No

For Product Safety Concerns and Information please contact our EU representative GPSR@taylorandfrancis.com Taylor & Francis Verlag GmbH, Kaufingerstraße 24, 80331 München, Germany

Printed and bound by CPI Group (UK) Ltd, Croydon, CR0 4YY
29/10/2025
01987646-0020